Modern Welding

Modern Welding

Summer Conroy

NY RESEARCH PRESS
New York

Published by NY Research Press
118-35 Queens Blvd., Suite 400,
Forest Hills, NY 11375, USA
www.nyresearchpress.com

Modern Welding
Summer Conroy

International Standard Book Number: 978-1-63238-861-2 (Hardback)

Cataloging-in-Publication Data

Modern welding / Summer Conroy.
p. cm.
Includes bibliographical references and index.
ISBN 978-1-63238-861-2
1. Welding. 2. Metal-work. 3. Ironwork. 4. Forging. I. Conroy, Summer.
TS227 .M63 2022
671.52--dc23

Contents

Preface

Welding is a sculptural or fabrication process where materials such as thermoplastics and metals are joined. The joining process makes use of heat to melt the parts. The subsequent cooling of these parts causes fusion. Different sources of energy can be used for welding such as electric arc, gas flame, laser and electron beam. Arc welding is a type of welding where electric current is used to maintain an electric arc between the base material and an electrode in order to melt the metal at the welding point. A few subtypes of arc welding are gas tungsten arc welding and gas metal arc welding. The topics included in this book on welding are of utmost significance and bound to provide incredible insights to readers. This book elucidates the concepts and innovative models around prospective developments with respect to welding. It strives to provide a fair idea about this discipline and to help develop a better understanding of the latest advances within this field.

A foreword of all Chapters of the book is provided below:

Chapter 1 - The sculptural process that deals with joining metals by using heat for melting and cooling for fusion of joints is referred to as welding. A few types of welding include gas welding, arc welding, resistance welding, radiant energy welding, etc. This is an introductory chapter which will briefly introduce all these types of welding as well as welding electrodes; **Chapter 2 -** The welding process that uses electricity for creating heat to melt and join metals is called arc welding. Some common types of arc welding are shielded metal arc welding, electric arc welding, gas tungsten welding, plasma arc welding. The topics elaborated in this chapter will help in gaining a better perspective about these different types of arc welding; **Chapter 3 -** Resistance welding is the type of welding process that makes use of electrical resistance to produce heat for forming the weld. There are various types of resistance welding such as resistance spot welding, flash butt welding, resistance seam welding, projection welding, etc. This chapter closely examines these different types of resistance welding to provide an extensive understanding of the subject; **Chapter 4 -** The welding process that joins the pieces of metals with the use of high pressure and a temperature below the melting point of the parent metal is defined as solid state welding. It includes explosive welding, forge welding, friction welding, solid state diffusion welding, etc. This chapter discusses in detail the concepts related to solid state welding; **Chapter 5 -** Welding metallurgy involves the chemical, mechanical, and physical properties of metals. Weldability is the ability of a material to be welded. This chapter has been carefully written to provide an easy understanding of welding metallurgy and weldability of different materials; **Chapter 6 -** Many other approaches come under the domain of welding and joining which can be named as brazing, soldering and adhesive bonding. This chapter closely examines these approaches of welding and joining as well as welding process for plastics to provide an extensive understanding of the subject.

I would like to thank the entire editorial team who made sincere efforts for this book and my family who supported me in my efforts of working on this book. I take this opportunity to thank all those who have been a guiding force throughout my life.

Summer Conroy

1

Introduction to Welding

The sculptural process that deals with joining metals by using heat for melting and cooling for fusion of joints is referred to as welding. A few types of welding include gas welding, arc welding, resistance welding, radiant energy welding, etc. This is an introductory chapter which will briefly introduce all these types of welding as well as welding electrodes.

Welding is a technique used for joining metallic parts usually through the application of heat. This technique was discovered during efforts to manipulate iron into useful shapes. Welded blades were developed in the 1st millennium CE, the most famous being those produced by Arab armourers at Damascus, Syria. The process of carburization of iron to produce hard steel was known at this time, but the resultant steel was very brittle. The welding technique—which involved interlayering relatively soft and tough iron with high-carbon material, followed by hammer forging—produced a strong, tough blade.

In modern times the improvement in iron-making techniques, especially the introduction of cast iron, restricted welding to the blacksmith and the jeweler. Other joining techniques, such as fastening by bolts or rivets, were widely applied to new products, from bridges and railway engines to kitchen utensils.

Modern fusion welding processes are an outgrowth of the need to obtain a continuous joint on large steel plates. Riveting had been shown to have disadvantages, especially for an enclosed container such as a boiler. Gas welding, arc welding and resistance welding all appeared at the end of the 19th century.

Basic Principles of Welding

A weld can be defined as a coalescence of metals produced by heating to a suitable temperature with or without the application of pressure, and with or without the use of a filler material.

In fusion welding a heat source generates sufficient heat to create and maintain a molten pool of metal of the required size. The heat may be supplied by electricity or by a gas flame. Electric resistance welding can be considered fusion welding because some molten metal is formed.

Solid-phase processes produce welds without melting the base material and without the addition of a filler metal. Pressure is always employed, and generally some heat is provided. Frictional heat is developed in ultrasonic and friction joining, and furnace heating is usually employed in diffusion bonding.

The electric arc used in welding is a high-current, low-voltage discharge generally in the range 10–2,000 amperes at 10–50 volts. An arc column is complex but, broadly speaking, consists of a cathode that emits electrons, a gas plasma for current conduction, and an anode region that becomes comparatively hotter than the cathode due to electron bombardment. A direct current (DC) arc is usually used, but alternating current (AC) arcs can be employed.

Total energy input in all welding processes exceeds that which is required to produce a joint, because not all the heat generated can be effectively utilized. Efficiencies vary from 60 to 90 percent, depending on the process; some special processes deviate widely from this figure. Heat is lost by conduction through the base metal and by radiation to the surroundings.

Most metals, when heated, react with the atmosphere or other nearby metals. These reactions can be extremely detrimental to the properties of a welded joint. Most metals, for example, rapidly oxidize when molten. A layer of oxide can prevent proper bonding of the metal. Molten-metal droplets coated with oxide become entrapped in the weld and make the joint brittle. Some valuable materials added for specific properties react so quickly on exposure to the air that the metal deposited does not have the same composition as it had initially. These problems have led to the use of fluxes and inert atmospheres.

In fusion welding the flux has a protective role in facilitating a controlled reaction of the metal and then preventing oxidation by forming a blanket over the molten material. Fluxes can be active and help in the process or inactive and simply protect the surfaces during joining.

Inert atmospheres play a protective role similar to that of fluxes. In gas-shielded metal-arc and gas-shielded tungsten-arc welding an inert gas—usually argon—flows from an annulus surrounding the torch in a continuous stream, displacing the air from around the arc. The gas does not chemically react with the metal but simply protects it from contact with the oxygen in the air.

The metallurgy of metal joining is important to the functional capabilities of the joint. The arc weld illustrates all the basic features of a joint. Three zones result from the passage of a welding arc: (1) the weld metal, or fusion zone, (2) the heat-affected zone, and (3) the unaffected zone. The weld metal is that portion of the joint that has been melted during welding. The heat-affected zone is a region adjacent to the weld metal that has not been welded but has undergone a change in microstructure or mechanical properties due to the heat of welding. The unaffected material is that which was not heated sufficiently to alter its properties.

Weld-metal composition and the conditions under which it freezes (solidifies) significantly affect the ability of the joint to meet service requirements. In arc welding, the weld metal comprises filler material plus the base metal that has melted. After the arc passes, rapid cooling of the weld metal occurs. A one-pass weld has a cast structure with columnar grains extending from the edge of the molten pool to the centre of the weld. In a multipass weld, this cast structure may be modified, depending on the particular metal that is being welded.

The base metal adjacent to the weld, or the heat-affected zone, is subjected to a range of temperature cycles, and its change in structure is directly related to the peak temperature at any given point, the time of exposure, and the cooling rates. The types of base metal can be grouped in three classes: (1) materials unaffected by welding heat, (2) materials hardened by structural change, (3) materials hardened by precipitation processes.

Welding produces stresses in materials. These forces are induced by contraction of the weld metal and by expansion and then contraction of the heat-affected zone. The unheated metal imposes a restraint on the above, and as contraction predominates, the weld metal cannot contract freely, and a stress is built up in the joint. This is generally known as residual stress, and for some critical applications must be removed by heat treatment of the whole fabrication. Residual stress is unavoidable in all welded structures, and if it is not controlled bowing or distortion of the weldment will take place. Control is exercised by welding technique, jigs and fixtures, fabrication procedures, and final heat treatment.

There are a wide variety of welding processes:

Forge Welding

This original fusion technique dates from the earliest uses of iron. The process was first employed to make small pieces of iron into larger useful pieces by joining them. The parts to be joined were first shaped, then heated to welding temperature in a forge and finally hammered or pressed together. The Damascus sword, for example, consisted of wrought-iron bars hammered until thin, doubled back on themselves, and then rehammered to produce a forged weld. The larger the number of times this process was repeated, the tougher the sword that was obtained. In the Middle Ages cannons were made by welding together several iron bands, and bolts tipped with steel fired from crossbows were fabricated by forge welding. Forge welding has mainly survived as a blacksmith's craft and is still used to some extent in chain making.

Arc Welding

Shielded metal-arc welding accounts for the largest total volume of welding today. In this process an electric arc is struck between the metallic electrode and the workpiece. Tiny globules of molten metal are transferred from the metal electrode to the weld joint. Since arc welding can be done with either alternating or direct current, some welding units accommodate both for wider application. A holder or clamping device with an insulated handle is used to conduct the welding current to the electrode. A return circuit to the power source is made by means of a clamp to the workpiece.

Gas-shielded arc welding, in which the arc is shielded from the air by an inert gas such as argon or helium, has become increasingly important because it can deposit more material at a higher efficiency and can be readily automated. The tungsten electrode version finds its major applications in highly alloyed sheet materials. Either direct or alternating current is used, and filler metal is added either hot or cold into the arc. Consumable electrode gas-metal arc welding with a carbon dioxide shielding gas is widely used for steel welding. Two processes known as spray arc and short-circuiting arc are utilized. Metal transfer is rapid, and the gas protection ensures a tough weld deposit.

Submerged arc welding is similar to the above except that the gas shield is replaced with a granulated mineral material as a flux, which is mounded around the electrode so that no arc is visible.

Plasma welding is an arc process in which hot plasma is the source of heat. It has some similarity to gas-shielded tungsten-arc welding, the main advantages being greater energy concentration, improved arc stability, and easier operator control. Better arc stability means less sensitivity to joint alignment and arc length variation. In most plasma welding equipment, a secondary arc must first be struck to create an ionized gas stream and permit the main arc to be started. This secondary arc may utilize either a high-frequency or a direct contact start. Water cooling is used because of the high energies forced through a small orifice. The process is amenable to mechanization, and rapid production rates are possible.

Thermochemical Processes

One such process is gas welding. It once ranked as equal in importance to the metal-arc welding processes but is now confined to a specialized area of sheet fabrication and is probably used as much by artists as in industry. Gas welding is a fusion process with heat supplied by burning acetylene in oxygen to provide an intense, closely controlled flame. Metal is added to the joint in the form of a cold filler wire. A neutral or reducing flame is generally desirable to prevent base-metal oxidation. By deft craftsmanship very good welds can be produced, but welding speeds are very low. Fluxes aid in preventing oxide contamination of the joint.

Another thermochemical process is aluminothermic (thermite) joining. It has been successfully used for both ferrous and nonferrous metals but is more frequently used for the former. A mixture of finely divided aluminum and iron oxide is ignited to produce a superheated liquid metal at about 2,800 °C (5,000 °F). The reaction is completed in 30 seconds to 2 minutes regardless of the size of the charge. The process is suited to joining sections with large, compact cross sections, such as rectangles and rounds. A mold is used to contain the liquid metal.

Resistance Welding

Spot, seam, and projection welding are resistance welding processes in which the required heat for joining is generated at the interface by the electrical resistance of the joint. Welds are made in a relatively short time (typically 0.2 seconds) using a low-voltage, high-current power source with force applied to the joint through two electrodes, one on each side. Spot welds are made at regular intervals on sheet metal that has an overlap. Joint strength depends on the number and size of the welds. Seam welding is a continuous process wherein the electric current is successively pulsed into the joint to form a series of overlapping spots or a continuous seam. This process is used to weld containers or structures where spot welding is insufficient. A projection weld is formed when one of the parts to be welded in the resistance machine has been dimpled or pressed to form a protuberance that is melted down during the weld cycle. The process allows a number of predetermined spots to be welded at one time. All of these processes are capable of very high rates of production with continuous quality control. The most modern equipment in resistance welding includes complete feedback control systems to self-correct any weld that does not meet the desired specifications.

Flash welding is a resistance welding process where parts to be joined are clamped, the ends brought together slowly and then drawn apart to cause an arc or flash. Flashing or arcing is continued until the entire area of the joint is heated; the parts are then forced together and pressure maintained until the joint is formed and cooled.

Low- and high-frequency resistance welding is used for the manufacture of tubing. The longitudinal joint in a tube is formed from metal squeezed into shape with edges abutted. Welding heat is governed by the current passing through the work and the speed at which the tube goes through the rolls. Welding speeds of 60 metres (200 feet) per minute are possible in this process.

Electron-beam Welding

In electron-beam welding, the workpiece is bombarded with a dense stream of high-velocity electrons. The energy of these electrons is converted to heat upon impact. A beam-focusing device is included, and the workpiece is usually placed in an evacuated chamber to allow uninterrupted electron travel. Heating is so intense that the beam almost instantaneously vaporizes a hole through the joint. Extremely narrow deep-penetration welds can be produced using very high voltages—up to 150 kilovolts. Workpieces are positioned accurately by an automatic traverse device; for example, a weld in material 13 mm (0.5 inch) thick would only be 1 mm (0.04 inch) wide. Typical welding speeds are 125 to 250 cm (50 to 100 inches) per minute.

Cold Welding

Cold welding, the joining of materials without the use of heat, can be accomplished simply by pressing them together. Surfaces have to be well prepared, and pressure sufficient to produce 35 to 90 percent deformation at the joint is necessary, depending on the material. Lapped joints in sheets and cold-butt welding of wires constitute the major applications of this technique. Pressure can be applied by punch presses, rolling stands, or pneumatic tooling. Pressures of 1,400,000 to 2,800,000 kilopascals (200,000 to 400,000 pounds per square inch) are needed to produce a joint in aluminum; almost all other metals need higher pressures.

Friction Welding

In friction welding two workpieces are brought together under load with one part rapidly revolving. Frictional heat is developed at the interface until the material becomes plastic, at which time the rotation is stopped and the load is increased to consolidate the joint. A strong joint result with the plastic deformation and in this sense the process may be considered a variation of pressure welding. The process is self-regulating, for, as the temperature at the joint rises, the friction coefficient is reduced and overheating cannot occur. The machines are almost like lathes in appearance. Speed, force, and time are the main variables. The process has been automated for the production of axle casings in the automotive industry.

Laser Welding

Laser welding is accomplished when the light energy emitted from a laser source is focused upon a workpiece to fuse materials together. The limited availability of lasers of sufficient power for most welding purposes has so far restricted its use in this area. Another difficulty is that the speed and the thickness that can be welded are controlled not so much by power but by the thermal conductivity of the metals and by the avoidance of metal vaporization at the surface. Particular applications of the process with very thin materials up to 0.5 mm (0.02 inch) have, however, been very successful. The process is useful in the joining of miniaturized electrical circuitry.

Diffusion Bonding

This type of bonding relies on the effect of applied pressure at an elevated temperature for an appreciable period of time. Generally, the pressure applied must be less than that necessary to cause 5 percent deformation so that the process can be applied to finished machine parts. The process has been used most extensively in the aerospace industries for joining materials and shapes that otherwise could not be made—for example, multiple-finned channels and honeycomb construction. Steel can be diffusion bonded at above 1,000 °C (1,800 °F) in a few minutes.

Ultrasonic Welding

Ultrasonic joining is achieved by clamping the two pieces to be welded between an anvil and a vibrating probe or sonotrode. The vibration raises the temperature at the interface and produces the weld. The main variables are the clamping force, power input, and welding time. A weld can be made in 0.005 second on thin wires and up to 1 second with material 1.3 mm (0.05 inch) thick. Spot welds and continuous seam welds are made with good reliability. Applications include extensive use on lead bonding to integrated circuitry, transistor canning, and aluminum can bodies.

Explosive Welding

Explosive welding takes place when two plates are impacted together under an explosive force at high velocity. The lower plate is laid on a firm surface, such as a heavier steel plate. The upper plate is placed carefully at an angle of approximately 5° to the lower plate with a sheet of explosive material on top. The charge is detonated from the hinge of the two plates, and a weld takes place in microseconds by very rapid plastic deformation of the material at the interface. A completed weld has the appearance of waves at the joint caused by a jetting action of metal between the plates.

Weldability of Metals

Carbon and low-alloy steels are by far the most widely used materials in welded construction. Carbon content largely determines the weldability of plain carbon steels; at above 0.3 percent carbon some precautions have to be taken to ensure a sound joint. Low-alloy steels are generally regarded as those having a total alloying content of less than 6 percent. There are many grades of steel available, and their relative weldability varies.

Aluminum and its alloys are also generally weldable. A very tenacious oxide film on aluminum tends to prevent good metal flow, however, and suitable fluxes are used for gas welding. Fusion welding is more effective with alternating current when using the gas-tungsten arc process to enable the oxide to be removed by the arc action.

Copper and its alloys are weldable but the high thermal conductivity of copper makes welding difficult. Refractory metals such as zirconium, niobium, molybdenum, tantalum, and tungsten are usually welded by the gas-tungsten arc process. Nickel is the most compatible material for joining, is weldable to it and is extensively used in dissimilar metal welding of steels, stainlesses and copper alloys.

Power Supply for Welding Processes

Selection of a welding process is determined primarily by the characteristics of the joint, the materials involved their shape and thickness, and joint design. Additionally, production requirements, such as rate and quality, must also be considered. Only after the process has been determined can the proper power supply and accessory equipment be chosen. The process is the primary factor in their selection.

Power and control requirements for these processes are somewhat more sophisticated than for conventional shielded metal arc or stick electrode welding. But since the average metallurgist or welding engineer isn't an electrical expert, the selection of proper power supply and the reasons for it can be confusing.

The basic types of power supplies are:

- Constant current.
- Constant voltage.

Constant Current Power Supply

The conventional stick electrode welder is sometimes called a 'constant current machine'. It is also called a 'dropper' because its voltage drops as welding current increases, thus its volt- ampere output curve 'droops'.

With the machine turned on but with no arc, and hence no current flowing, it has a relatively high open circuit voltage of 70-80 volts. Generally speaking welding is done at the steeper portions of the curve and this is ideal for manual stick electrode welding.

Arc voltage depends upon the physical length of the arc between the electrode and the work and this can never be held completely constant in manual welding. But, since rate of burn off of filler metal is determined by the amount of current, burn off stays substantially constant if current doesn't vary. There are many variations of this type of machine based upon power input (single or three phase), output (ac, dc, or ac/dc), and the type of output control (mechanical or electrical).

Constant Voltage Power Supply

The other basic type of arc welding power supply produces a constant voltage. Thus at any voltage setting current may vary from zero to an extremely high short circuit current. Such a machine is designed specifically for gas shielded metal arc welding and is not generally suitable for stick electrode welding.

Actually, no welding machine can produce a truly constant voltage. In practice, voltage drops at least 1 volt for each 100 amp output. Nevertheless, short circuit currents may be as high as several thousand amperes.

Normally, constant voltage machines have lower open circuit voltages than the constant current machines, about 50 volts maximum as compared to 80 volts. As a means of obtaining the desired arc voltage, the operator sets open circuit voltage, rather than current, at the machine. Settings may range from 10 to 48 volts.

Welding current can reach several thousand amperes at short circuit. Current adjusts itself to burn off filler metal at a rate sufficient to maintain the arc length required by the present voltage and is thus determined by the rate of electrode feed.

Welding Electrode

A welder needs an electrode to generate an electric current to do arc welding. In welding, an electric current is conducted through an electrode which is used to join the parent metals. When you keep electrode tip near the parent metal electric current jumps from the electrode tip to the parent metal. The main purpose of electrodes used in welding is to create an electric arc. These electrodes can be positively charged anode or they can be negatively charged cathode.

Factors that you must consider before selecting welding electrodes:

- The electrode rod should have greater tensile strength than the parent metals.
- You have to consider joint design, shape, specifications of base metals and welding positions.

Types of Welding Electrodes

Basically, depending upon the process there are two types of welding electrodes:

- Consumable Electrodes.
- Non-Consumable Electrodes.

Consumable Electrodes

Consumable electrodes have low melting point. These types of welding electrodes are preferred to use in Metal Inert Gas (MIG) welding. For making consumable electrodes, materials such as mild steel and nickel steel are used. The one precaution that you must take is to replace consumable electrodes after regular intervals. The only disadvantage of using such electrodes is that they don't have a large number of industry applications but at the same time they are easy to use and maintain.

Consumable electrodes are categorized as:

- Bare Electrodes: Bare electrodes are electrodes without any type of coating and mostly used in applications where there is no need of coated electrode.
- Coated Electrodes: Coated electrodes are classified according to the coating factor. Coating factor is the ratio of the diameter of the electrode to the diameter of the core wire.

So, following are sub types of coated electrodes:

- Light coated electrodes with coating factor of 1.25. Light coating applied to electrodes helps to remove impurities such as oxides and phosphorous. Light coating also helps in enhancing arc stability.
- Medium coated electrodes with coating factor of 1.45.
- Shielded arc or heavily coated electrodes with coating factor ranging between 1.6 to 2.2. These electrodes have a proper and well defined composition. The heavily coated electrodes are designed in three types – electrodes with cellulose coating, electrodes with mineral coating and the electrodes with coating of both cellulose as well as mineral coating.

Non-consumable Electrodes

These types of welding electrodes are also referred to as Refractory electrodes. There are again two sub-types of non-consumable electrodes:

- Carbon or Graphite electrodes: It is made up of carbon and graphite and mostly used in the applications of cutting and arc welding.
- Tungsten electrodes: Basically, it is consists of tungsten as the name itself suggests and it is a non-filler metal electrode.

As the name suggests, these types of welding electrodes are not consumed in the entire welding process or we can say more appropriately that they do not melt during welding. But practically, due to the vaporization and oxidation processes taking place during welding there is a little bit reduction in the length of the electrode. The non-consumable electrodes have high melting point and are unable to fill the gap in the workpiece. Non-consumable electrodes are made from materials such as pure tungsten, graphite or carbon coated with copper. The melting point of carbon is 3350 °C and that of Tungsten is 3422 °C. Non-consumable electrodes are used in Tungsten inert gas welding (TIG) and carbon arc welding.

Important Characteristics of Non-Consumable Electrodes:

- While using non-consumable electrodes you have to use shielding gases. The shielding gases are inert gases and the reason behind to use them is to protect welding area from oxygen and surrounding atmosphere.

- The non-consumable electrodes are usually made as cathode and the workpieces as anode.

Classification of Welding Electrodes

This classification is based on practical approach towards using and selecting welding electrodes. The American Welding Society has classified electrodes in different formats for proper understanding of different electrodes easily and to identify them comfortably.

Suppose we consider the electrode named as E6018-X:

- Here the E indicates that this is an electrode.
- The consecutive two digits after the letter E gives you the tensile strength of the electrode. This tensile strength is measured in psi and this strength is 1000 times the given number. That means here this tensile strength of the given electrode is 60,000 psi.
- Here 1 indicates the welding position. The welding position is indicated by 1,2 and 4:
    - 1 indicates flat, horizontal, vertical position.
    - 2 indicates flat, horizontal position.
    - 4 indicates flat, horizontal, vertically downward position.
- The number 8 gives you information about the type of coating and current used. It also tells about the penetration of electrode means the electrode may penetrate deep, low, medium.
- The X in the E6018-X tells us about the additional specifications of an electrode. Here the term X is not always mentioned. It is used only when an electrode has some additional features. This classification is applicable to mild steel coated electrode. If you consider other types of electrodes, the classification will be same but only the feature represented by X letter may vary.
- Some additional properties represented by letter X:
    - -1: It indicates that the electrode is more ductile and has high toughness.
    - -M: It is comfortable for military applications and low moisture content.
    - -H4, -H8,-H16: All represents maximum diffusible hydrogen limit measured in millimetres per 100 Grams. For instance, -H4 = 4 mL per 100 grams.

Precautions that you must take while handling welding electrodes:

- You must always keep electrodes dry.
- As moisture destroys electrode coating and is very harmful for electrodes. So, as soon as electrodes get dried, you must keep welding electrodes in moisture free environment. There are various containers available which give you moisture free experience.
- Last but not the least, never bend the electrodes as bending can harm electrode coating.

2

Arc Welding

The welding process that uses electricity for creating heat to melt and join metals is called arc welding. Some common types of arc welding are shielded metal arc welding, electric arc welding, gas tungsten welding, plasma arc welding. The topics elaborated in this chapter will help in gaining a better perspective about these different types of arc welding.

Arc welding is a welding process, in which heat is generated by an electric arc struck between an electrode and the work piece.

Electric arc is luminous electrical discharge between two electrodes through ionized gas.

Any arc welding method is based on an electric circuit consisting of the following parts:

- Power supply (AC or DC),
- Welding electrode,
- Work piece,
- Welding leads (electric cables) connecting the electrode and work piece to the power supply.

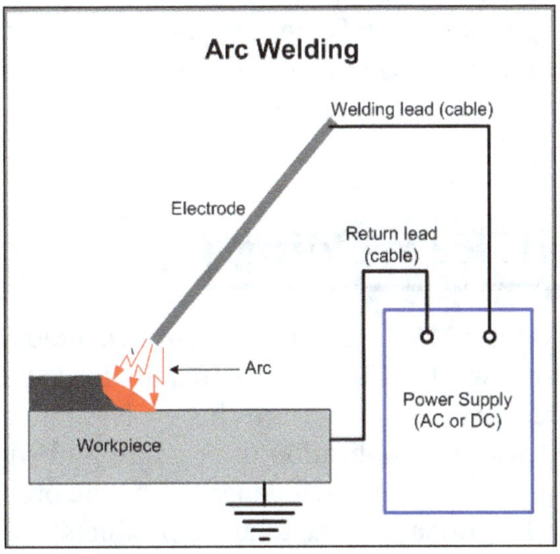

Electric arc between the electrode and work piece closes the electric circuit. The arc temperature may reach 10000 °F (5500 °C), which is sufficient for fusion the work piece edges and joining them.

When a long join is required the arc is moved along the joint line. The front edge of the weld pool melts the welded surfaces when the rear edge of the weld pool solidifies forming the joint.

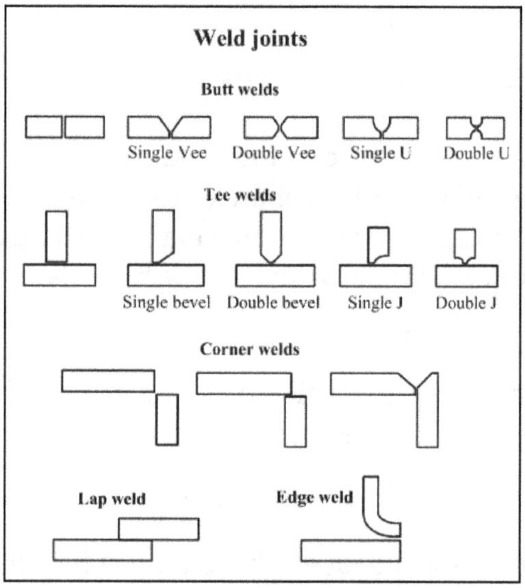

Figure: Types of weld joints.

When a filler metal is required for better bonding, filling rod (wire) is used either as outside material fed to the arc region or as consumable welding electrode, which melts and fills the weld pool. Chemical compositions of filler metal is similar to that of work piece.

Molten metal in the weld pool is chemically active and it reacts with the surrounding atmosphere. As a result weld may be contaminated by oxide and nitride inclusions deteriorating its mechanical properties. Neutral shielding gases (argon, helium) and shielding fluxes are used for protection of the weld pool from atmospheric contamination. Shields are supplied to the weld zone in form of a flux coating of the electrode or in other forms.

Shielded Metal Arc Welding

In shielded metal arc welding (SMAW), an arc between a covered electrode and a weld pool is used to accomplish a weld. As the welder steadily feeds the covered electrode into the weld pool, the decomposition of the covering evolves into gases that shield the pool. The process is used without application of pressure and with filler metal from the covered electrode. The sound weld metal deposited by the process is used for both joining and for applying a functional surface to metal products.

Because of the many possible variations in the composition of the electrode covering and the large selection of core wire chemistry, the process can produce an extensive range of weld metal deposits with desirable mechanical and physical properties, while providing for a smooth arc, uniform metal transfer characteristics, and ease of operation. It is one of the oldest and simplest welding processes and continues to be widely used.

The simplicity of the process extends to the number and nature of the circuit components required, which are as follows:

- A power source of adequate current rating and duty cycle.
- A SMAW electrode compatible with the output of the power source.
- A suitably sized welding cable.
- An electrode holder.
- A workpiece lead.

Process Fundamentals

The distinctive feature of SMAW is the physical presence of the covering/coating that surrounds the core wire of the consumable electrode. The covered rod is called an electrode because it functions as the terminal from which the electric flow changes from the conducting solid to the conducting plasma of the welding arc.

For any given application, the electrode must meet the following criteria:

- It must shield the arc and the weld metal.
- It must add metal to the weld.
- It must sustain a welding arc.

The constituents of the covering accomplish these functions. When sufficiently heated, the covering's ingredients do the following:

- Decompose into gases and displace the air at the weld site, thus providing a shield for the arc and the weld metal.
- Ionize to support the arc plasma.
- Flux the molten metal and, on cooling, form a protective slag cover on the weld bead.

The covering may also contain metal powders that enhance the metal contribution of the electrode to the weld pool.

Principles of Operation

The SMAW process uses an electric circuit that supports a welding arc to convert electric line power or fuel into heat. The heat from the welding arc is intense and extremely concentrated. It immediately melts a portion of the workpiece and the end of the electrode. The welder maintains the arc length by holding a consistent space between the electrode and the weld pool that forms on the workpiece. When the arc is removed, the liquid fuses and the melt solidifies into continuous metal.

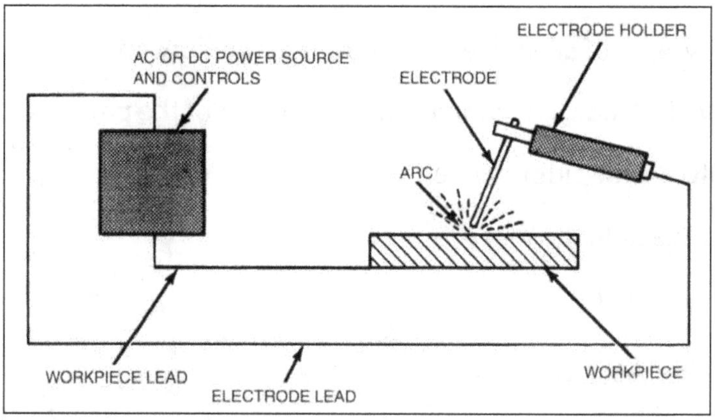

Figure: Elements of a typical welding circuit for shielded metal arc welding.

As shown in the schematic in figure, the power source is connected into a circuit with the electrode and workpiece in series. The welding cable used in the circuit, the electrode holder, and the connection between the cable and the workpiece are also important elements of the circuit. The power source has two distinct output terminals. From one terminal, a connection is made to the electrode. When using direct current (DC), the proper terminal for the electrode connection is determined by the required polarity for that type electrode. When using alternating current (AC), the electrode may be connected to either terminal. The circuit is open between the workpiece and the electrode.

As long as the SMAW electrode is held away from the workpiece, the circuit remains open and a voltmeter can be used to measure the voltage drop between the electrode holder and the workpiece for this open-circuit (prewelding) condition.

Covered Electrodes

All SMAW electrodes have a covering with constituents that facilitate the welding process and add alloying elements that impart useful properties to the weld. Without the covering, the arc would be very difficult to maintain, the weld deposit would be brittle with dissolved oxygen and nitrogen, the weld bead would be dull and irregularly shaped, and the workpiece would be undercut.

Electrode manufacturers apply the covering on SMAW electrodes either by extrusion or dipping. Extrusion is much more widely used and is achieved by mixing the dry components with liquid silicates. The dipping process is employed primarily for SMAW electrodes used to weld cast iron and for some specialty electrodes that have a complex core wire.

The covering contains most of the stabilizing, shielding, fluxing, deoxidizing, and slag-forming materials essential to the process. Besides sustaining the arc and supplying filler metal for the weld deposit, decomposition of the electrode covering introduces other key materials into or around the arc, or both. Depending on the type of electrode being used, the electrode covering provides the following:

- A gas to shield the arc and prevent excessive atmospheric contamination of the molten metal.
- Deoxidizers to react with and deplete the level of dissolved gaseous elements that can cause porosity.
- Fluxing agents to accelerate chemical reactions and cleanse the weld pool.
- A slag blanket to protect the hot weld metal from the air and to enhance the mechanical properties, bead shape, and surface cleanliness of the weld metal.
- Alloying elements to achieve the desired microstructure.
- Elements and compounds to control grain growth.
- Alloying materials to improve the mechanical properties of the weld metal.
- Elements to affect the shape of the weld pool.
- Elements that affect the wetting of the workpiece and the viscosity of the liquid weld metal.
- Stabilizers to help establish the desirable electrical characteristics of the electrode and minimize spattering.

The chemical compounds in the covering, in combination with the core wire composition, create unique mechanical properties in the weld and enhance welding characteristics such as arc stability, metal transfer type, and slag. The different types of electrodes are formulated not only to weld different metals but also to optimize certain characteristics of the process and gain an advantage in a particular area of application.

Arc Shielding

The shielding action of the process is essentially the same for all SMAW electrodes, but the specific method of shielding and the volume of slag produced vary from one electrode type to another.

Two mechanisms are at work to prevent the detrimental effect on the weld pool caused by the gases contained in the air. The first is the forceful displacement of the air by gases produced by the burning and decomposition of the electrode covering. The second is the blanketing action of the flux or slag, which prevents diffusion of the air constituents into the liquid metal. Electrode coverings vary in their reliance on these two mechanisms to provide the most advantageous shielding action for a specific weld.

Process Advantages

A main advantage of SMAW is the large variety of metals and alloys the process is capable of welding. Procedures and electrodes are available to weld carbon and low-alloy steels, high-alloy steels, coated steels, tool and die steels, stainless and heat-resisting steels, cast irons, copper and copper alloys, plus nickel and cobalt alloys.

Following are other advantages of the process:

- The equipment is relatively simple, inexpensive, and portable.
- The SMAW electrode provides both the shielding and the filler metal to make a sound weld.
- Auxiliary gas shielding or granular flux is not required.
- The process is less sensitive to wind and draft than the gas shielded arc welding processes.
- The dimensions of the SMAW electrodes are ideal for reaching into areas of limited access (electrodes can be bent, and with the aid of mirrors, applied in blind spots).
- The process is suitable for most of the commonly used metals and alloys.
- The process is flexible and can be applied to a variety of joint configurations and welding positions.
- Optimum results can be readily and reliably obtained.

Process Limitations

- Metals with low melting temperatures, such as lead, tin, and zinc, and their alloys are not welded with SMAW. These metals have relatively low boiling points and the intense heat of the SMAW arc immediately cause them to vaporize from the solid state. Shielded metal arc welding is likewise not suitable for reactive metals such as titanium, zirconium, tantalum, and niobium because the shielding provided is not sufficiently inert to prevent contamination of the weld.
- The process yields lower deposition rates than the gas metal arc welding (GMAW) and flux cored arc welding (FCAW) processes. The deposition rate is lower because the maximum useful current is limited.
- If the electrode is too long or if the current is too high, the amount of heat generated within the SMAW electrode will be excessive. After welding has started, the

temperature of the covering will eventually rise to a range that will cause premature breakdown of the covering. That breakdown, in turn, triggers a deterioration of the arc characteristics and reduces the level of shielding. Consequently, welding must stop before the electrode has been fully consumed. Hence, the amount of current that can be used is limited within a range that prevents overheating of the electrode and breakdown of the covering. The limited useful current results in generally lower deposition rates than those obtainable with GMAW or FCAW.

- Stub loss is another inherent drawback. The stub is the grip end of the SMAW electrode that is discarded. It consists of the core wire within the grip of the electrode holder and a small portion of the covered length. Stub loss affects deposition efficiency, not the deposition rate. Longer stub losses translate directly into lower deposition efficiency.
- The operator factor — arc time as a percentage of the welder's total labor time — for SMAW is usually lower than that obtained with a continuous electrode process such as GMAW or FCAW.

When the weldment requires a large volume of filler metal, the combination of low deposition rates and a lower operator factor detracts from the use of SMAW. In these instances, the weld completion rate may be too slow and the weld cost relatively high.

Power Sources for Shielded Metal Arc Welding

The first part of being successful at stick welding or shielded metal arc welding is to make sure that we have a power source that is capable of doing the work. To perform shielded metal arc welding we need to have a constant current power supply. This is a type of machine that produces welding power that has a decrease in voltage as the amperage increases. Some people call this a drooping slope machine, in reference to the volt-amp curve produced by charting the machine output on a graph.

One way to identify a constant current power source is when the main knob on the face of the machine controls amperage. If you see no knob on the machine to adjust amperage but you can find one that adjusts voltage, you got yourself a constant voltage power source. A constant voltage power source is generally used for wirefeed processes like MIG and flux core arc welding. We can make MIG welds with the constant current power source provided we have the proper voltage sensing wire feeder, like the LN-25 from Lincoln Electric.

Understanding that we need a constant current power source for stick welding now we can consider what type of variations are available for constant current power source.

An engine driven machine is one that is powered by either a gasoline or diesel internal combustion engine. The engine rotated generator produces the welding power. These types of machine are great when we do not have local power from our utility. This is the type of machine that we want if we are going to do stick welding or shielded metal arc welding in a remote location, possibly away from workshop. The engine generally runs in an idle mode and then ramps up and increases RPM when we strike the arc and start welding.

Unlike other welding machines that plug into the wall, this type of machine requires additional maintenance to care for the engine, that's the heart of the power source. Also, this is not the type of machine we want to use if we have to weld and the power source needs to be located indoors. The exhaust from the engine will spoil the breathing and local atmosphere rapidly. If we have the ability to plug into the wall and get power from the local utility our options are little bit more varied.

The most basic type of welding machine powered by electricity is a type called the transformer. A transformer machine takes a high-voltage low amperage power from the utility and transforms it into high current low-voltage power we use for welding. This machine generally will have a large transformer inside and is also larger and heavier than a comparable inverter machine. These types of machines generally produce only AC power.

There is another type of machine that is very closely related to the transformer, the transformer-rectifier machine. A transformer-rectifier machine is a transformer machine that has a rectifier circuit installed in it to allow the machine to produce DC power from the AC input received from the utility. Depending upon how we have the cables configured to the output lugs on the front of the machine we can produce welding power in the DC mode either with the electrode positive or electrode negative. Which polarity we might select will depend upon the job we're doing the position were working in the electrodes we have selected to use.

Inverters seem to be the current trend in the welding industry. Inverter machines eliminate the large heavy transformer and replace it with electronic circuitry and printed circuit boards. Most inverter machines are capable of producing both AC and DC power. One characteristic of the inverter machines are that their generally smaller in size and lighter in weight than the comparable transformer counterpart. Inverter machines are also capable of changing the waveform and the frequency of the output power in ways that the transformer machines are not capable of doing.

Electric Arc Welding

Electric Arc Welding is a welding process which is used for welding the metals with the help of electricity to generate sufficient heat for softening the metal, as well as when the softened metal is cooled then the metals will be welded. This kind of welding uses a power supply to make an arc among a metal stick & the base material to soften the metals at the end of the contact.

These welders can utilize either DC otherwise AC, & electrodes like consumable otherwise non-consumable. Generally, the welding location can be defended with some kind of shielding gas, slag, otherwise, vapor. This welding process could be manual, fully or semi-automated.

By using Electric Arc, If the heat required for melting of plates is obtained, called as Electric Arc Welding Process. The fusion of metals by an electric arc is one of the most important processes in the industry. This is commonly called arc welding or SMAW (Shielded Metal Arc Welding).

Applications

Arc welding is mostly used in the manufacturing and construction industries.

Example: Pipe Welding, Metal Joining for rotary Kilns, etc.

Process and Equipment

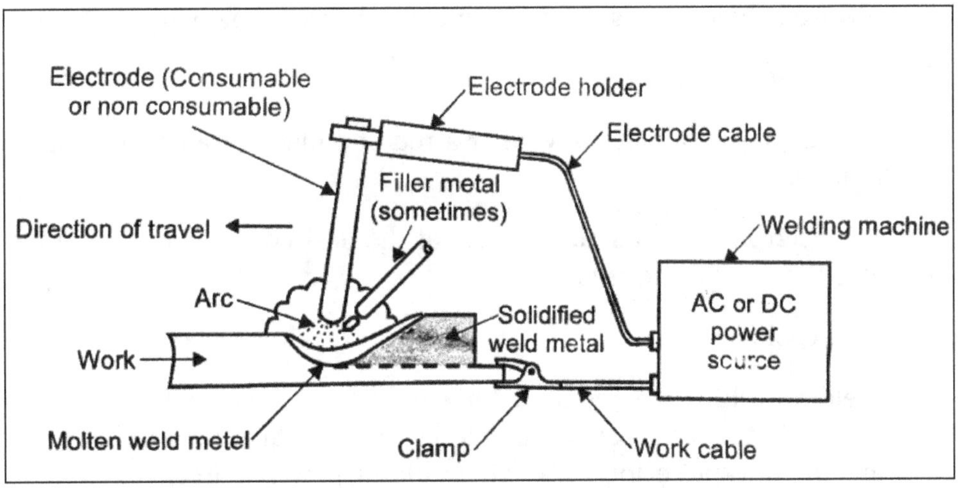

Figure: Electric Arc Welding Process.

The Electric Arc Welding Process Equipment generally consists of the following parts:

- Arc welding power source,
- Welding Electrodes,
- Welding cables,
- Electrode holder,
- Hand Screen,
- Chipping Hammer,
- Wire brush,

- Protective clothing.
- Arc welding power source:
    - Both direct current (DC) and alternating current (AC) are used for electric arc welding, each having its particular applications.
    - For AC welding supply, transformers are predominantly used for almost all Arc welding processes.
    - DC welding supply is usually obtained from generators driven by the electric motors.
    - They have to step down the usual supply voltage (200-400 volts) to the normally open circuit welding voltage (50-90 volts) by means of Step-down Transformer.
- Electrode holder:
    - Electrode holder is used for holding the electrode manually.
- Welding Electrodes:
    - An electrode is a piece of wire or a rod of a metal or alloy, with or without coatings.
    - An Electrode can generate the arc at the desired welded location between electrode and workpiece.
- Welding cables:
    - Welding cables are required for conduction of current from the power source to various parts of Arc welding process equipment i.e. electrode, the arc, the workpiece and back to the welding power source.
    - These are insulated copper or aluminum cables.
- Hand Screen:
    - Hand screen is used for the protection of eyes during Arc welding process.
- Chipping hammer:
    - It is used for striking the slag from the weld bead region.
- Wire brush:
    - A wire brush is generally used to clean the surface before and after the welding process.

- Protective clothing:
  - The operator wears protective clothing such as apron to keep away from the exposure of direct heat to the body.
  - Always wear gloves before touching any item in the workshop.

Principle

- The work to be welded is connected to one side of an electric circuit (Anode), and a metal electrode is connected to the other side (Cathode).
- When the power supply is given and the optimum gap is maintained between Cathode and Anode, the very high velocity negatively charged electrons will be generated at the cathode.
- These will be attracted by the anode and moving towards the anode. When this very high velocity negatively charged electrons are impinging on to the anode, the Kinetic Energy of electrons is converted into Heat energy.
  - Therefore, the heat generation is taking place at the anode.
  - Simultaneously, the positively charged ions also will be generated at the anode which is attracted by the cathode and moving towards the cathode.
  - When these high-velocity ions impinging on to the cathode, the Kinetic Energy of ions is converted into Heat energy.
  - Therefore, heat generation is taking place at cathode also.

The ratio of heat generation between the cathode and anode is given as follows:

Heat Generation of Cathode: Anode=2:1.

Spark Zone in Electric Arc Welding Process

- During the flow of electrons and ions, some of them may get collided in between the path so that the Kinetic Energy possessed by both the elements is converted into heat energy.
- This spontaneous release of heat energy will induce the sparkling and in the spark zone, the temperatures induced are very high which is about 5000 °C-6000 °C.
- This high-temperature zone will produce UV rays. Hence the Electric Arc Welding process Zone must be seen only through safety goggles to safeguard the human eye.

Power Supply for Electric Arc Welding

The power supply can use either direct (DC) or alternating (AC) current. The type of current used can sometimes cause problems in welding. Any electrical current is either a direct current (DC) or an alternating current (AC). AC and DC are terms that refer to the polarity of the electrical current that the welding machine creates and runs through the electrode.

Direct current (DC) is electricity flowing in a constant direction and possessing a voltage with constant polarity, either positive or negative. Typical DC currents include the current in batteries and often used in low voltage devices such as cell phones and remote controls.

In welding, electrode negative (DC) or straight current provides faster deposition rates because there is quicker melt-off of the electrode. Electrode positive (DC) results in deeper penetration.

Alternating current (AC) is electricity that switches direction back and forth so the voltage also periodically reverses because the current changes direction. Typical AC currents are what you would expect to see from your electrical outlets in your home and often used in higher voltage devices such as household appliances.

AC current changes its polarity 120 times per second with a 60-hertz current. Reversed polarity (AC) results in deeper penetration. In Alternate Current (AC) welding, since the current and the magnetic field of the arc reverse direction many times a second, there is no net deflection of the arc.

Applications of AC Welding

AC welding can be used to weld magnetic metals.

This cannot be done with DC welding. AC welding is ideal for the following types of welds:

- Downhand heavy plate,
- Fast fill,
- Aluminum TIG welding with high frequency.

The primary advantage of using AC welding is that it allows the weld operator to weld on magnetized materials.

In AC current, the current changes direction and is not affected by magnetism. The arc remains stable and is easier to control.

AC welding is also the preferred method for:

- TIG welding aluminium, because the current supports welding at a higher temperature.
- Making repairs on machinery because the machinery usually has a magnetized field and is older and may have rusty areas where there is concern about the higher heat penetration that can occur with DC welding.
- Seam welding in shipbuilding because the current settings can often be higher than those used in DC welding and a deeper penetration of plate metals can be obtained.

The biggest drawback to using AC welding is the quality of the weld. It is not as smooth as DC welding because of the continuous change in directional flow and there is likely to be more spatters.

Applications of DC Welding

Welding with DC is best used for:

- Hard facing,
- Overhead or vertical welding,
- Build-up of heavy deposits,
- Single carbon brazing,
- Stainless steel TIG welding,
- Cutting tap.

As a rule of thumb, DC is preferred for welding because:

- It produces a smoother weld and there are fewer spatters because of the constant linear direction of the current.
- It maintains a constant and stable arc and is thus is easier to handle and more reliable than AC current.
- Machinery that uses DC current is generally cheaper and easier to use.
- It welds thinner metals better than AC current.

The primary drawbacks to using of DC welding are:

- There's greater potential for arc blow.
- DC currents are not supplied by electrical grids, so they need an internal transformer to change the current from AC to C for use. That makes this method more expensive to use.
- It's not good for use in welding aluminum because it cannot produce the high-intensity heat necessary.

Arc Blow Issues

In welding, there can be a lot of reasons for a weld failure. One of the most frustrating is magnetic arc blow. This occurs when a magnetic field is present in the area where the weld is being prepared to join two metal components.

The quality of the weld can be affected by magnetic disturbance due to the interaction of the magnetic field of the welding current with the residual magnetic field present in the metal.

The disturbance causes the welding arc to deflect away from the weld path and behave erratically.

The welding operator must fight arc blow by trying to keep the electrical current stable and the work usually suffers in quality. Arc blow occurs only in direct current (DC) welding. An effective fix for the problem is to switch from DC current to AC current.

Overall Strength of Welds

Overall the strength of the weld can be determined by many factors, such as:

- Proper electrode, welding apparatus and procedures.
- Properties of the materials being welded-magnetic vs. non-magnetic.

- Proper edge preparation-the cleaner the edge, the better the weld.
- Current settings - DC vs. AC.
- Speed of travel - the angle of the electrode needs to be maintained throughout the length of the joint as it is being welded.

A strong weld can be achieved in both AC welding and DC welding so long as you weld with the current and polarity appropriate for the material being welded.

Gas Tungsten Arc Welding

The gas tungsten arc welding (GTAW) is also known as TIG welding (Tungsten inert gas). The consumable tungsten electrode is produced the arc to generate the required heat. It is similar to form of arc welding process. In this method the arc generated between the base metal and tungsten electrode. When the electrode is used to only create the arc and it is non-consumed in welding process. The welding gun provided inside of electrode holder. It gun made of ceramics or it is cooled by using water.

The weld zone, electrode and arc are surrounded by an inert gas. Thus gas usually as helium, argon or both are mixed. The argon is mostly used in welding process because of it is heavier and produce better shielded low flow rating.

The welding gun hold the electrode, carries inert gas and electric current. When the electric power is supplied by electric transformer through hoses. In this gas flow rate controlled by the value fitted on the cylinder. Strikes the arc by maintain the air gap between electrode and work piece, the gas inserted from the gun for the value in the torch.

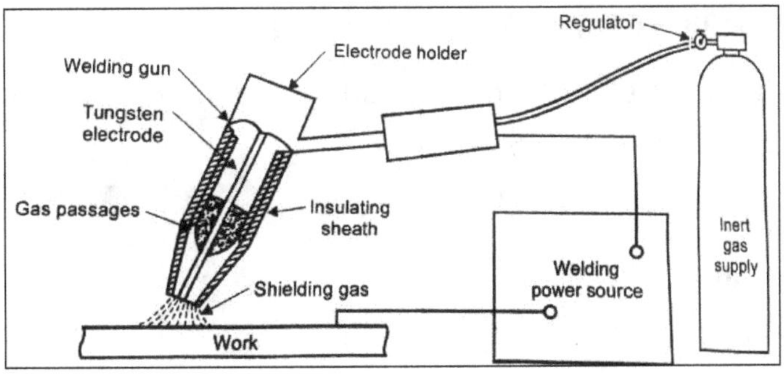

Figure: Gas tungsten arc welding.

Now the welding area displaces the shielding gas to atmospheric air and the contamination of weld metal with hydrogen and oxygen in the air. If the filler wire are provided between the molten materials. For high speed welding process the automatic filler wire feed with rated speed into the arc is used. The filler wire is similarly in composition to welded material.

In this gas tungsten arc welding process the AC power is used for best of non-ferrous material. The DC power source is suitable for ferrous alloys in straight polarities. The straight polarities will be reducing the loss of tungsten electrode.

The electrodes are alloyed with thorium (around 1%) or zirconium. For alternating current application, Zr alloyed tungsten is used. Pure Ti electrodes are suitable for AC welding of aluminum and magnesium.

The arc welding from voltage may be range from 10v to 15v with current for 50-350 amps in argon material and voltage of 15-25v with current of 50-350 amps in helium shielding gas. It TIG welding is suitable for 7 mm metal thickness and below.

Advantages of Gas Tungsten Arc Welding

Some advantages of gas tungsten arc welding:

- All position of metal will be welded.
- The weld is equal to base metal in composition.
- The finished weld surfaces not require cleaning process because of flux is not used.
- Smoke or fumes not presented to obscure vision.
- The heat is concentrated in the small area, so distortion of base metal is minimal.
- The metal is not transferred across the arc, so no splatter is provided.
- Thus produce high quality weld on non-ferrous metal.

Application of Gas Tungsten Arc Welding

It is general to common welding process of aluminium. The gas tungsten arc welding used to weld for following materials:

- Copper base alloys,
- Nickel base alloys,
- Low alloy steel,
- Dissimilar metal like as copper,
- Titanium and its alloy,
- Hard falling,
- Magnesium alloy.

It is suitable for thin sheet 0.5 to 3 mm thickness produce weld with excellent quality and surface finish. It is used for pressure vessels, pipes, and heat exchanger where tight less is important.

Tungsten Electrodes

The electrode material for GTAW is made from a tungsten alloy. Tungsten has one of the highest melting temperatures of any metal, about 6,170 °F (3,410 °C).

The size of an electrode to be used is determined by the welding current required. Larger electrodes permit higher currents to be used. Smaller diameter electrodes may be used for welding thinner materials or while welding out of position.

Following is a list of various types of tungsten alloys used:

- Pure tungsten is used on nonferrous metals, such as aluminum and magnesium, and is typically used with a balled-end preparation on alternating current (AC).
- Thoriated tungsten is the most common type of tungsten electrode for use on carbon and stainless steel. It can be purchased with 1 or 2 percent thorium. The thoriated tungsten starts readily and maintains a stable arc. It has a greater resistance to contamination and will maintain a sharp point and will not break down as readily as pure tungsten.

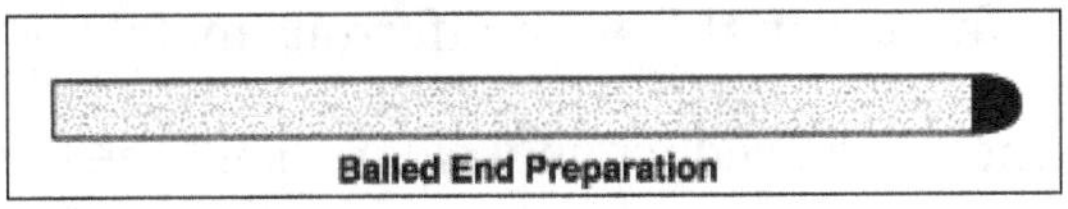

Figure: Balled End Preparation.

- Zirconiated tungsten is typically used for welding with higher AC currents on nonferrous metals.

Preparing a point or using an electrode taper angle applies to thoriated tungsten. Thoriated tungsten electrodes are ground to a point to give better arc starting, with high frequency added. This provides the arc ignition and keeps the electrode from contacting the work. It also aids in stabilizing the arc.

The degree of taper affects the shape and amount of penetration of a weld. To reduce the number of times the electrode must be sharpened, the welder needs to develop the skill of not touching the tungsten to the work during the welding process. The recommended taper length is 21/2 to 3 times the diameter of the electrode.

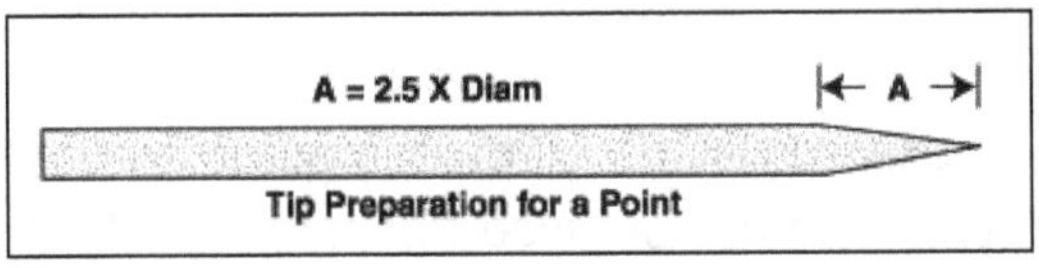

Figure: Tip Preparation for a point.

Shielding Gases

Argon and helium are the two most commonly used shielding gases used for GTAW. The characteristics most desirable for shielding purposes are the chemical inertness of the gases and their ability to produce smooth arc action at high currents. Both gases are inert, causing an ionization effect in the welding arc. They protect the tungsten electrode and the molten weld pool from the atmosphere.

Gas purity affects a weld. Metals will withstand small amounts of impurities, but, for the best results, the percent of inert gas used should be at least 99.9 percent pure.

Argon is heavier than helium and may be supplied in liquid or gaseous form. Argon provides for good cleaning action. The flow rates are determined by the size of the tungsten and the gas cup diameter. Argon is suitable for welding similar and dissimilar metals and works well while welding in the vertical and overhead welding positions.

Helium is a lighter inert gas. It can be distributed as a liquid, but is used more often as a compressed gas. It leaves the weld area faster than argon, and higher flow rates are necessary when using it. Helium produces a narrow but deep heat-affected zone (HAZ), which is good for welding on heavier metals. It is suitable for welding at high speeds and gives good coverage in vertical and overhead welding positions. It helps to increase the penetration, and when used as a back purge, it tends to flatten the pass of the weld bead. Helium is suitable for use on thicker nonferrous metals.

Argon and helium mixtures are used when welders need the control of the argon and the penetration of the helium. This mixture is not necessary when welding plain carbon steels. Typical mixtures vary, depending on the application. It is often used for automatic welding applications.

Argon and hydrogen mixtures are often used for welding of stainless steel, INCONEL, and MONEL. This mixture should not be used when welding plain carbon steels. The typical mixture is a 95 percent argon and 5 percent hydrogen. Nitrogen can also be used as a shielding gas, but is rarely used because of its higher current requirements. It is suitable for welding copper.

Welding Current and Joint Design

The current depends primarily on the type of metal to be welded, the current levels required, and the availability of the machine that produces that type of welding current.

Direct current electrode positive (DCEP) (reverse polarity) is sometimes used to weld very thin nonferrous metals, and is also used for balling the tungsten electrode. Direct current electrode negative (DCEN) (straight polarity) is used most commonly to weld stainless steel and ferrous metals.

AC current, with the addition of high frequency, is most commonly used for welding some nonferrous metals such as aluminum and magnesium. It provides good cleaning action and gives moderate penetration.

Weld Joint Design

The five basic types of joints are the butt joint, the corner joint, the edge joint, the lap joint, and the tee joint. Of the five types of joint designs, the butt and the tee joint are the most commonly used. The strength of a weld joint is another factor contributing to weld joint design. Weld joints can be either partial or full penetration, depending on the strength required of the joint. Weld joint design or weldment configuration for GTAW is determined by the type of metal, configuration of the weldment, designated codes and specifications, and the metallurgical analysis. Several factors influence the joint design to be used, including the strength required, the welding position, the metal thickness, and how accessible the joint is to the welder.

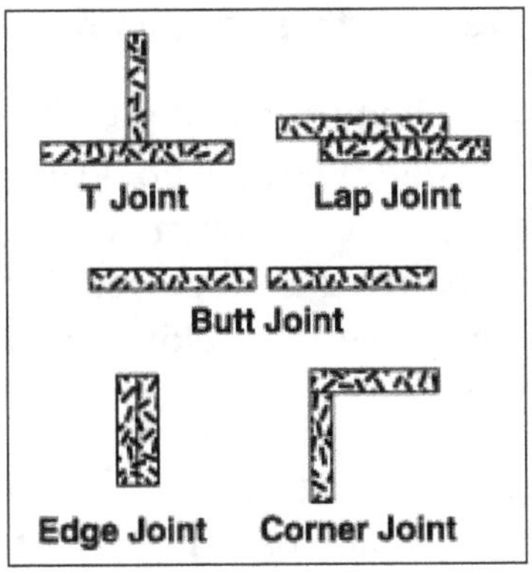

The purpose of any joint design is to produce a sound weld deposit with the desired properties as economically as possible. The edge and joint preparation are important because they will affect both the quality and the cost of welding.

Preweld Preparations

Before using GTAW, several steps must be taken to prepare the electrode and the weld joint, fixture the weldment, set the variables and preheat the base metal, if necessary. The amount of preparation depends on the size of the weldment, type of base material, fit-up, and the quality requirements.

- Electrode Preparation: Electrode preparation depends on the type of electrode and the welding application. The tip may have a ground point or a ball end for welding with AC.

  To prepare an electrode with a point, the grind marks should run parallel to the electrode. To prepare a ball on the end of tungsten, the power supply must be switched to DCEP (reverse polarity). Then, after starting the arc between the electrode and a piece of scrap metal or copper, it must be maintained at a moderate current level. The tip of the ball should be perfectly clean, shiny, and have a mirror-like finish.

- Preparing the Weld Joint: When preparing the weld joint, several different methods can be used, including oxyfuel cutting, plasma cutting, shearing, machining, air carbon arc gouging, grinding, or chipping. Remember, preparing the weld joint properly will help produce a sound weldment and meet the requirements of quality standards for welding.

- Cleaning: Cleaning the material to be welded is important. GTAW welds are often susceptible to contamination during welding. The surface to be welded must be free from oil, grease, paint, dirt, oxides, and other foreign material.

  Aluminum has an oxide coating that, if not removed, will contaminate the weld area. Cleaning solutions, wire brushes, grinders, and abrasive blasting are some of the methods used to remove these contaminants.

- Fixturing and Positioning: Fixturing and positioning will also affect the shape, size, and uniformity of a weld. Fixtures hold the weldment in place while controlling distortion, helping to locate and maintain parts in their position relative to the weldment. When fixturing is employed, it can reduce the time for welding. Positioning will help move the weldment into a flat position to improve productivity for the welder.

  Chill blocks, heat sinks, or backing bars may be used when welding some metals to prevent burn-through, reduce base material temperatures, or to minimize distortion.

- Preheating: Depending on the alloying elements in the base material, the thickness of the steel, and the configuration of the joint, preheat is sometimes needed. The amount of preheat needed for a given application is usually obtained from the welding procedure. Several methods of controlling preheat temperature are available, including furnace heating, electric induction coils, oxyfuel torches, and resistance heating blankets.

  The preheat temperature can be measured using temperature sticks, crayon pellets, temperature indicators, thermocouples, thermistors, or infrared thermometers.

Selecting a GTAW Power Source

One of the benefits of gas tungsten arc welding (GTAW) is its ability to control amperage, or heat input, with a high degree of precision, especially when using it in mechanized applications.

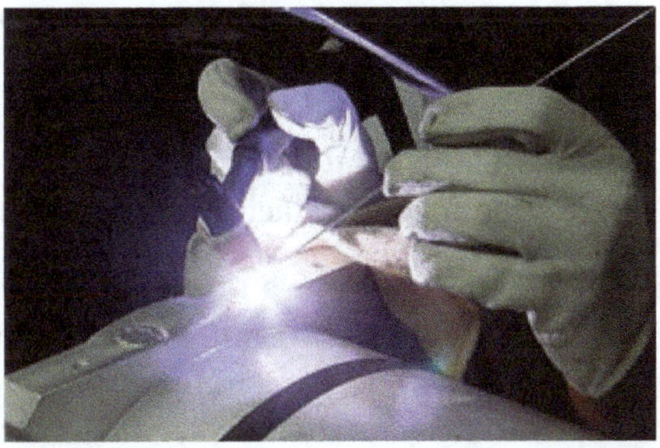

A key element of GTAW is the power source, which generates the arc. Choosing the right power source can be a confusing decision. Consider several factors before purchasing a GTAW power source:

- The application,
- Material to be welded,
- Power requirements,
- Usage,
- Time constraints,
- Additional capacity.

GTAW uses a non-consumable tungsten-alloy electrode. It's held in a torch assembly, which initiates and maintains the arc. In manual GTAW, the heat of the arc commonly is controlled one of two ways: through the use of a foot pedal or as a static setting on the power source. The foot pedal provides variable control of the welding amperage, or heat.

The torch assembly is either water- or air-cooled. A solenoid on the power source activates the flow of shielding gas to the torch. Some power sources provide an additional solenoid, which switches the water flow to the torch, while others don't have a solenoid for the torch coolant so the cooler can circulate coolant constantly through the torch, providing cooling all the time.

High frequency (HF) may be required to start the arc for non-touch-start applications. Most GTAW power sources provide HF capability, but not all do.

A GTAW power source is a constant-current machine. Unlike constant-voltage power sources, it will do its best to maintain a constant current (i.e., heat) despite changes in the voltage (i.e., arc length). A constant-voltage power source does just the opposite, which is required for gas metal arc welding (GMAW). The most important consideration when deciding to purchase a power source is the intended application of the process. One size does not fit all, and if you purchase the wrong machine for the job, then nothing but problems and frustration will result.

When choosing a GTAW power source, consider first output current and duty cycle. The output current of the power source is one of the primary factors, because it dictates the thickness of the materials that can be welded. This is a complicated measurement that requires you to understand your application in terms of material type, thermal conductivity, and part size. These are all contributing factors that affect amperage requirements.

Duty cycle is a rating of the amount of power a machine can produce in a given time. The duty cycle rating is broken down on a 10-minute scale. For example, a machine with a duty cycle rating of 150 amps at 60 percent duty cycle means that the machine can produce 150 amps for six minutes and then must cool for four minutes. A machine with a rating of 300 amps at 100 percent duty cycle means that the machine can produce 300 amps of current continuously.

The combination of output current and duty cycle is crucial. If you choose a power source with a lot of output current and a poor duty cycle, for instance, you'll be able to weld at higher amperages, but for short periods of time. For welding at home this may not be a problem, but when time is critical, waiting for a machine to cool isn't cool.

For productivity and safety, regulating input and output power is essential. This often is overlooked until the machine is hooked up and running. Serious electrical problems can occur in a home-based application if the machine is not installed properly. If you don't have a comprehensive understanding of electricity, consult a professional before attempting to wire or install power to a welding machine.

However, the type of input power also should be a deciding factor when choosing a GTAW power source. Some machines can operate at multiple-line input voltages, for example, 208 V, 220 V, and 440 V. As a rule, the higher the input voltage, the less current (amperage) the machine will draw through your electrical system. In a shop where several power sources operate, this is a critical factor.

Newer Power Sources

Newer inverter-based machines offer additional flexibility. This is especially apparent in the use of AC for welding nonferrous materials. Some manufacturers offer power sources with a maximum control over the arc waveform. Some of these newer machines also are considered to be multiprocess power sources, which differ from conventional power sources because they are not limited to either a constant-current or constant-voltage mode.

Inverter-based machines can offer efficiency in terms of power input versus output, which translates into more work with less energy, thus reducing cost and space. These machines are powerful for their size, but also cost more.

While some newer machines offer technology and flexibility, achieving high-quality welds with a less advanced, less expensive power source is possible. Technology can't beat what a skilled welder can provide. Many machines are affordable and can provide years of service.

A plug-and-chug equation doesn't exist for buying a power source. However, a comprehensive understanding of the process and the materials to be welded is imperative and can lead to a satisfying purchase.

Plasma Arc Welding

The Plasma arc welding (PAW) is similar to GTAW or gas tungsten welding. In this kind of welding process, the arc will generate among work part as well as the tungsten electrode. The major dissimilarity among plasma arc welding and gas tungsten welding is that the electrode is located within the torch of Plasma arc welding. It can be heated the gas at the temperature of 30000 °F & changes it into the plasma to attack the welding region.

Principle

The hot ionized gases are known as plasma. When a sufficient amount of energy provided to any inert gas, some of its electrons breaks free from its nucleus but travel with it. After the electrons leave, the atoms are converted into hot ionized state. It is most common state of matter which is known as fourth state of matter. These ionized atoms have high heat contain which is further used to join two plates. This is basic principle of plasma arc welding. This welding is extended form of TIG welding in which, a non-consumable tungsten electrode is used to produce arc. This arc heats up the inert gases which are provided from inner orifice around tungsten electrode. The heating temperature is about 30000 degree centigrade at which the gas converts into ionized form. This hot ionized gas further used to create a welding joint by fusion.

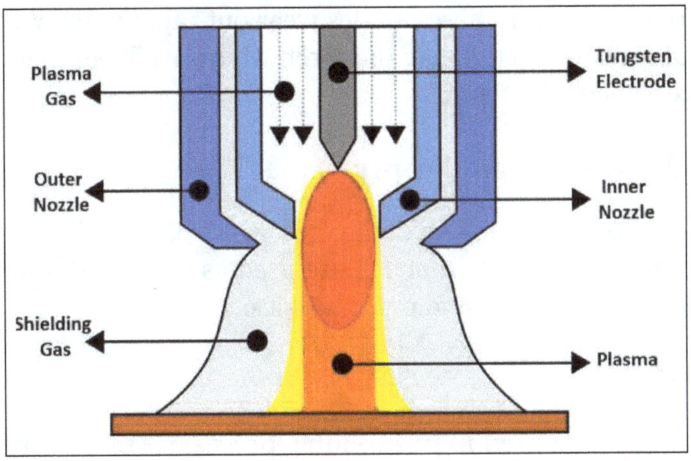

Figure: Principle of Plasma Arc Welding.

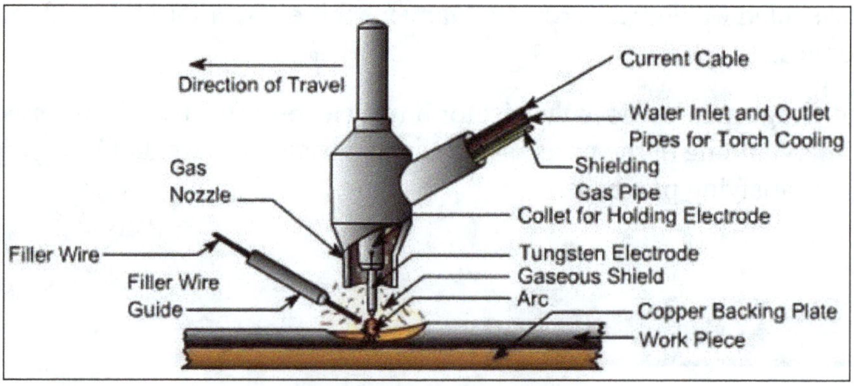

Figure: Plasma Arc Welding.

Power Source

PAW process needed a high power DC supply to generate electric spark in between tungsten electrode and welding plates (For transferred PAW Process) or in between tungsten electrode and discharge nozzle (For Non-transferred PAW process). One of the main features of plasma arc welding which make it different from TIG is its low ampere welding capability. This welding can weld at low ampere of about 2 amp and maximum current which it can handle is about 300 amp. It needs about 80 volts for proper working. The power source consist transformer, rectifier and control console.

Plasma Arc Torch

This is most important part of PAW process. This torch is quite similar as used in TIG welding but too complex. It consist four main parts which are tungsten electrode, collets, inner nozzle, and outer nozzle. The tungsten electrode is hold by the collet. The collet is available in varying diameters. The inner gas nozzle supply inert gases inside the torch to form plasma. The outside nozzle supply shielding gases which protect the weld area from oxidation. These nozzles wear out rapidly. PAW torches are water cooled because arc is contained inside the torch which produces high heat, so a water jacket is provided outside the torch.

Shielding and Plasma Gas Supply

Generally, plasma gas is same as shielding gas which is supplied by a same source. Mainly inert gases like argon, Helium etc. are used as both inert and shielding gases. This gas is supplied at both inert and outer nozzles.

Filler Material

Mostly no filler material is used in this welding process. If filler material is used, it is directly feed into weld zone.

Types

Mainly plasma arc welding can be classified into two types:

- Non-transferred Plasma Arc Welding: In this welding process, straight polarity DC current is used. In this process, the tungsten electrode is connected to the negative and the nozzle is connected to the positive pole. The arc produces between tungsten electrode and nozzle inside the torch. This will increase the ionization of gas inside the torch. The torch transfers this ionized gas for further process. It is used to weld thin sheets.
- Transferred Plasma Arc Welding: This process also uses straight polarity DC current. In this process, the tungsten electrode is connected to the negative terminal and the work piece is connected to the positive terminal. The arc is produces between tungsten electrode and work piece. In this process both plasma and arc transferred to the work piece which increases the heating capacity of process. It is used to weld thick sheets.

Working

This welding works on same as TIG instead, plasma is used to heat up the parent material. Its working can be summarized as follow:

- First the work pieces are properly cleaned. The power source supply power which produces arc between tungsten electrode and nozzle, or tungsten electrode and workpiece.
- The tungsten electrode gives a high intense arc which is used to ionization of gas particles and converts orifice gases into plasma. This hot ionized gas is supplied to the welding plates from a small orifice.
- The shielding gases like argon etc. are supplied through pressure valve and regulating valve to the outer nozzle of welding torch. These gases create a shield around the welding area which protects it from atmospheric gases like oxygen, nitrogen etc.
- The plasma strikes the welding plates and fuses it into one piece. Next the welding torch is moved in the direction of welding.
- If the welding required filler material, it is fed by the welder manually.

This is whole working process of plasma arc welding.

Application

- This welding is used in marine and aerospace industries.
- It is used to weld pipes and tubes of stainless steel or titanium.

- It is mostly used in electronic industries.
- It is used to repair tools, die and mold.
- It is used to welding or coating on turbine blade.

Advantages and Disadvantages

Advantages:

- High welding speed.
- High energy available for welding. It can be easily used to weld hard and thick work pieces.
- The distance between tool and work piece does not effects the arc formation.
- Low power consumption for same size weld.
- More stable arc produced by PAW method.
- High intense arc or high penetration rate.
- It can work at low amperage.

Disadvantages:

- Higher equipment cost.
- Noisy operation.
- More radiation.
- High skill labor required.
- High maintenance cost.

Carbon Arc Welding

Carbon arc welding is a process in which the joining of metals is produced by heating with an arc between a carbon electrode and the work. No shielding is used. Pressure and filler metal may or may not be used.

Carbon Arc Welding Equipment

- Electrodes: Carbon arc welding electrodes range in size from 1/8 to 7/8 in. (3.2 to 22.2 mm) in diameter. Baked carbon electrodes last longer than graphite electrodes. Figure shows typical air-cooled carbon electrode holders. Water-cooled holders are available for use with the larger size electrodes, or adapters can be fitted to regular holders to permit accommodation of the larger electrodes.

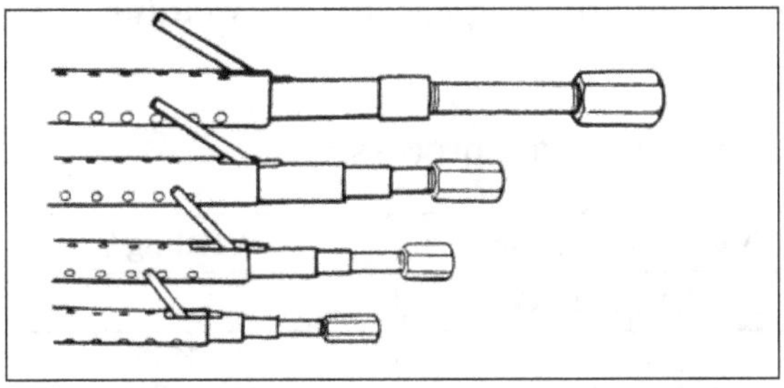

Figure: Typical air cooled carbon electrode holders.

- Machines: Direct current welding machines of either the rotating or rectifier type are power sources for the carbon arc welding process.
- Welding circuit and welding current: The welding circuit for carbon arc welding is the same as for shielded metal arc welding. The difference in the apparatus is a special type of electrode holder used only for holding carbon electrodes. This type of holder is used because the carbon electrodes become extremely hot in use, and the conventional electrode holder will not efficiently hold and transmit current to the carbon electrode. The power source is the conventional or constant current type with drooping volt-amp characteristics. Normally, a 60 percent duty cycle power source is utilized. The power source should have a voltage rating of 50 volts, since this voltage is used when welding copper with the carbon arc.

  Single electrode carbon arc welding is always used with direct current electrode negative (DCEN), or straight polarity. In the carbon steel arc, the positive pole (anode) is the pole of maximum heat. If the electrode were positive, the carbon electrode would erode very rapidly because of the higher heat, and would cause black carbon smoke and excess carbon, which could be absorbed by the weld metal. Alternating current is not recommended for single-electrode carbon arc welding. The electrode should be adjusted often to compensate for the erosion of carbon. From 3.0 to 5.0 in. (76.2 to 127.0 mm) of the carbon electrode should protrude through the holder towards the arc.

Advantages and Major Uses

The single electrode carbon arc welding process is no longer widely used. It is used for welding copper, since it can be used at high currents to develop the high heat usually required. It is also used for making bronze repairs on cast iron parts. When welding thinner materials, the process is used for making autogenous welds, or welds without added filler metal. Carbon arc welding is also used for joining galvanized steel. In this case, the bronze filler rod is added by placing it between the arc and the base metal.

The carbon arc welding process has been used almost entirely by the manual method of applying. It is an all-position welding process. Carbon arc welding is primarily used as a heat source to generate the weld puddle which can be carried in any position. Table shows the normal method of applying carbon-arc welding. Table shows the welding position capabilities.

Table: Method of applying carbon arc process.

Method of applying	Rating
Manual (MA)	A
Semiautomatic (SA)	No
Machine (ME)	B
Automatic (AU)	No

Table: Welding position capabilities.

Welding position	Rating
Flat	A
Horizontal FILLET	A
Horizontal	A
Vertical	A
Overhead	A
Pipe – fixed	--

- Weldable Metals: Since the carbon arc is used primarily as a heat source to generate a welding puddle, it can be used on metals that are not affected by carbon pickup or by the carbon monoxide or carbon dioxide arc atmosphere. It can be used for welding steels and nonferrous metals, and for surfacing.
- Steels: The main use of carbon arc welding of steel is making edge welds without the addition of filler metal. This is done mainly in thin gauge sheet metal work, such as tanks, where the edges of the work are fitted closely together and fused using an appropriate flux. Galvanized steel can be braze welded with the carbon arc. A bronze welding rod is used. The arc is directed on the rod so that the galvanizing is not burned off the steel sheet. The arc should be started on the welding rod or a starting block. Low current, a short arc length, and. rapid travel speed should be used. The welding rod should melt and wet the galvanized steel.
- Cast iron: Iron castings may be welded with the carbon arc and a cast iron welding rod. The casting should be preheated to about 1200 °F (649 °C) and slowly cooled if a machinable weld is desired.
- Copper: Straight polarity should always be used for carbon arc welding of copper. Reverse polarity will produce carbon deposits on the work that inhibit fusion. The work should be preheated in the range of 300 to 1200 °F (149

to 649 °C) depending upon the thickness of the parts. If this is impractical, the arc should be used to locally preheat the weld area. The high thermal conductivity of copper causes heat to be conducted away from the point of welding so rapidly that it is difficult to maintain welding heat without preheating. A root opening of 1/8 in. (3.2 mm) is recommended. Best results are obtained at high travel speeds with the arc length directed on the welding rod. A long arc length should be used to permit carbon from the electrode to combine with oxygen to form carbon dioxide, which will provide some shielding of weld metal.

Principles of Operation

Carbon arc welding, as shown in figure, uses a single electrode with the arc between it and the base metal. It is the oldest arc process, and is not popular today.

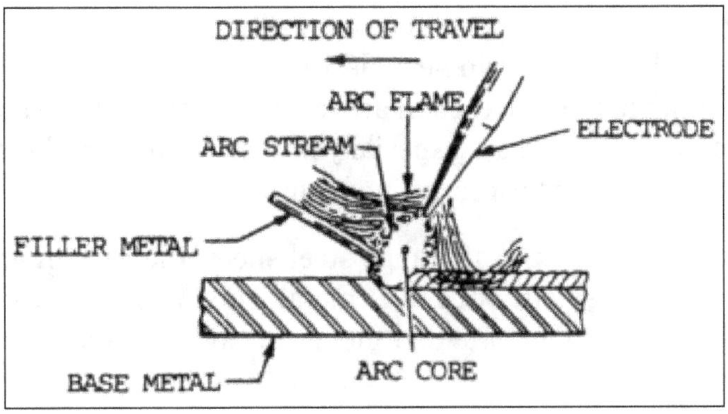

Figure: Process diagram – CAW.

In carbon arc welding, the arc heat between the carbon electrode and the work melts the base metal and, when required, also melts the filler rod. As the molten metal solidifies, a weld is produced. The non-consumable graphite electrode erodes rapidly and, in disintegrating, produces a shielding atmosphere of carbon monoxide and carbon dioxide gas. These gases partially displace air from the arc atmosphere and prohibit the oxygen and nitrogen from coming in contact with molten metal. Filler metal, when used, is of the same composition as the base metal. Bronze filler metal can be used for brazing and braze welding.

The workpieces must be free from grease, oil, scale, paint, and other foreign matter. The two pieces should be clamped tightly together with no root opening. They may be tack welded together.

Carbon electrodes 1/8 to 5/16 in. (3.2 to 7.9 mm) in diameter may be used, depending upon the current required for welding. The end of the electrode should be prepared with a long taper to a point. The diameter of the point should be about half that of the electrode. For steel, the electrode should protrude about 4.0 to 5.0 in. (101.6 to 127.0 mm) from the electrode holder.

A carbon arc may be struck by bringing the tip of the electrode into contact with the work and immediately withdrawing it to the correct length for welding. In general, an arc length between 1/4 and 3/8 in. (6.4 and 9.5 mm) is best. If the arc length is too short, there is likely to be excessive carburization of the molten metal resulting in a brittle weld.

When the arc is broken for any reason, it should not be restarted directly upon the hot weld metal. This could cause a hard spot in the weld at the point of contact. The arc should be started on cold metal to one side of the joint, and then quickly returned to the point where welding is to be resumed.

When the joint requires filler metal, the welding rod is fed into the molten weld pool with one hand while the arc is manipulated with the other. The arc is directed on the surface of the work and gradually moved along the joint, constantly maintaining a molten pool into which the welding rod is added in the same manner as in gas tungsten arc welding. Progress along the weld joint and the addition of a welding rod must be timed to provide the size and shape of weld bead desired. Welding vertically or overhead with the carbon arc is difficult because carbon arc welding is essentially a puddling process. The weld joint should be backed up, especially in the case of thin sheets, to support the molten weld pools and prevent excessive melt-thru.

For outside corner welds in 14 to 18 gauge steel sheet, the carbon arc can be used to weld the two sheets together without a filler metal. Such welds are usually smother and more economical to make than shielded metal arc welds made under similar conditions.

The welding schedule for carbon arc welding galvanized iron using silicon bronze filler metal is given in table. A short arc should be used to avoid damaging the galvanizing. The arc must be directed on the filler wire which will melt and flow on to the joint. For welding copper, use a high arc voltage and follow the schedule given in Table shows the welding current to be used for each size of the two types of carbon electrodes.

Table: Welding procedure schedule – Galvanized steel- Braze welding.

Material	Thickness		Electrode Size		Filler rod Size.		Welding	Arc voltage electrode
Gauge	In.	mm	In.	mm	In	mm	Amps dc	Neg
24	0.024	0.6	3/16	4.8	3/32	2.4	25-30	13-15
22	0.030	0.8	3/16	4.8	3/32	2.4	25-30	13-15
20	0.036	0.9	3/16	4.8	3/32	2.4	30-35	14-16
18	0.048	1.2	1/4	6.4	1/8	3.2	30-35	14-16
16	0.060	1.5	1/4	6.4	1/8	3.2	30-35	14-16
14	0.075	1.9	1/4	6.4	1/8	3.2	30-35	14-16
12	0.105	2.7	1/4	6.4	1/8	3.2	35-40	15-17

Table: Welding Procedure Schedule for Carbon Arc Welding Copper.

Thickness of copper			Diameter of Electrode and Filler Rod				Welding Current dc Amps	Voltage Electrode Negative
Decimal Inches	Fraction Inches on US Gauge		Electrode Carbon		Filler Rod			
	In.	Mm	In.	mm	In.	mm		
0.05	18				3/32		80	
0.0563	17		3/16	4.8	3/32	2.4	90	35
0.0625	1/16	1.6			1/8		90	
0.07	15		3/16	4.8	1/8	3.2	100	40
0.078	5/64	2.0			5/32	4.0	120	
0.094	3/32	2.4	1/4	6.4	5/32	4.0	135	
0.109	7/64	2.8			5/32	4.0	140	40
0.125	1/8	3.2			3/16		150	
0.141	9/64	3.6			3/16		160	
0.156	5/32	4.0	1/4	6.1	3/16		165	
0.172	11/64	4.4			3/16		170	
0.1875	3/16	4.8			3/16	4.8	185	45
0.203	13/64	5.2			1/4		200	
0.219	7/32	5.6			1/4		200	
0.234	15/64	6.0	5/16	7.9	1/4		205	
0.25	1/4	6.4			1/4		215	
0.266	17/64	6.7			1/4	6.4	225	45
0.281	9/32	7.1			5/16		250	
0.3125	5/16	7.9			5/16		250	
0.344	11/32	8.7	5/16	7.9	5/16		255	
0.375	3/8	9.5			5/16		270	
0.406	13/32	10.3			5/16	7.9	290	50
0.4375	7/16	11.1			3/8		300	
0.4688	15/32	11.9	3/8	9.5	3/8		310	
0.5	1/2	12.7			3/8	9.5	325	50

Table: Welding current for carbon electrode types.

Electrode diameter		Carbon electrodes DCEN Amps	Graphics electrodes DCEN Amps
in.	mm		
1/8	3.2	15-30	15-35
3/16	4.8	25-55	25-60
1/4	6.4	50-85	50-90
5/16	7.9	75-115	80-125

3/8	9.5	100-165	110-165
7/16	11.1	125-185	140-210
1/2	12.7	150-225	170-260
5/8	15.9	200-310	230-370
3/4	19.0	250-400	290-490
7/8	22.2	300-500	400-750

Variations of the Process

There are two important variations of carbon arc welding. One is twin carbon arc welding. The other is carbon arc cutting and gouging.

Twin carbon arc welding is an arc welding process in which the joining of metals is produced, using a special electrode holder, by heating with an electric arc maintained between two carbon electrodes. Filler metal may or may not he used. The process can also be used for brazing.

The twin carbon electrode holder is designed so that one electrode is movable and can be touched against the other to initiate the arc. The carbon electrodes are held in the holder by means of set screws and are adjusted so they protrude equally from the clamping jaws. When the two carbon electrodes are brought together, the arc is struck and established between them. The angle of the electrodes provides an arc that forms in front of the apex angle and fans out as a soft source of concentrated heat or arc flame. It is softer than that of the single carbon arc. The temperature of this arc flame is between 8000 and 9000 °F (4427 and 4982 °C).

Alternating current is used for the twin carbon welding arc. With alternating current, the electrodes will burn off or disintegrate at equal rates. Direct current power can be used, but when it is, the electrode connected to the positive terminal should be one size larger than the electrode connected to the negative terminal to ensure even disintegration of the carbon electrodes. The arc gap or spacing between the two electrodes most is adjusted more or less continuously to provide the fan shape arc.

The twin carbon arc can be used for many applications in addition to welding, brazing, and soldering. It can be used as a heat source to bend or form metal. The welding current settings or schedules for different size of electrodes are shown in table. The twin carbon electrode method is relatively slow and does not have much use as an industrial welding process.

Table: Welding current for carbon electrode (twin torch).

Carbon electrode diameter		Welding current amperes ac	Arc voltage	Base metal thickness	
In.	Mm			In.	Mm
1/4	6.4	55	35-40	1/16	1.6
5/16	7.9	75	35-40	1/8	3.2

3/8	9.5	95	35-40	1/4	6.4
3/8	9.5	120	35-40	Over 1/4	Over 6.4

Gas Metal Arc Welding

The Gas metal arc welding (GMAW) is also called as Metal inert gas welding (MIGW). It uses a fresh metal electrode which is protected by the gas like helium, argon, etc. These gases will protect the join area from oxidation and generates multiple welding material layers. In this type of arc welding process, a filler wire can be fed constantly using a non-consumable metal electrode for welding the metal.

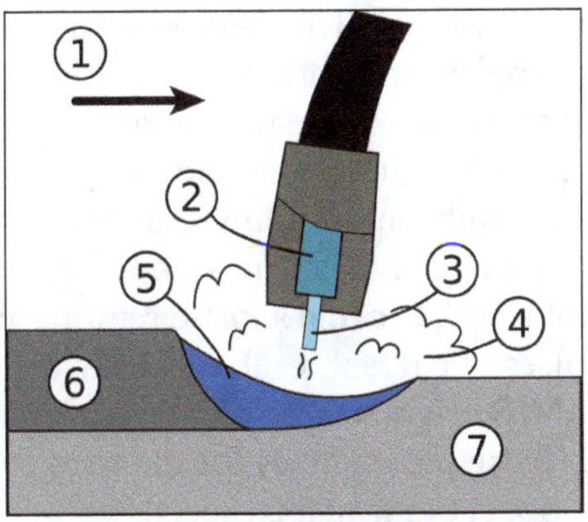

Figure: (1) Direction of travel, (2) Contact tube, (3) Electrode, (4) Shielding gas, (5) Molten weld metal, (6) Solidified weld metal, (7) Workpiece.

Process Characteristics

MIG is similar to MMA in that heat for welding is produced by forming an arc between a metal electrode and the workpiece; the electrode melts to form the weld bead. The main differences are that the metal electrode is a small diameter wire fed from a spool and an externally supplied shielding gas is necessary. As the wire is continuously fed, the process is often referred to as semi-automatic welding.

Metal Transfer Mode

The manner, or mode, in which the metal transfers from the electrode to the weld pool largely determines the operating features of the process. There are three principal metal transfer modes:

- Short circuiting,
- Droplet/spray,
- Pulsed.

Short-circuiting and pulsed metal transfer are used for low current operation while spray metal transfer is only used with high welding currents. In short-circuiting or "dip" transfer, the molten metal forming on the tip of the wire is transferred by the wire dipping into the weld pool. This is achieved by setting a low voltage; for a 1.2mm diameter wire, arc voltage varies from about 17V (100A) to 22V (200A). Care in setting the voltage and the inductance in relation to the wire feed speed is essential to minimise spatter. Inductance is used to control the surge in current which occurs when the wire dips into the weld pool.

For droplet or spray transfer, a much higher voltage is necessary to ensure that the wire does not make contact i.e. short-circuit, with the weld pool; for a 1.2mm diameter wire, the arc voltage varies from approximately 27V (250A) to 35V (400A). The molten metal at the tip of the wire transfers to the weld pool in the form of a spray of small droplets (about the diameter of the wire and smaller). However, there is a minimum current level, threshold, below which droplets are not forcibly projected across the arc. If an open arc technique is attempted much below the threshold current level, the low arc forces would be insufficient to prevent large droplets forming at the tip of the wire. These droplets would transfer erratically across the arc under normal gravitational forces. The pulsed mode was developed as a means of stabilising the open arc at low current levels i.e. below the threshold level, to avoid short-circuiting and spatter. Metal transfer is achieved by applying pulses of current, each pulse having sufficient force to detach a droplet. Synergic pulsed MIG refers to a special type of controller which enables the power source to be tuned (pulse parameters) for the wire composition and diameter, and the pulse frequency to be set the wire feed speed.

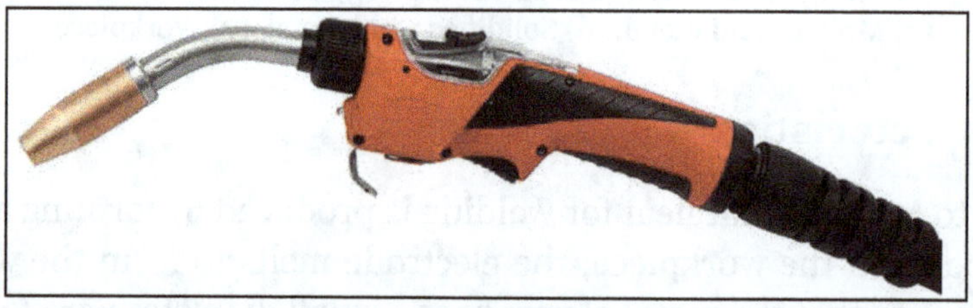

Figure: Typical MIG torch.

Shielding Gas

In addition to general shielding of the arc and the weld pool, the shielding gas performs a number of important functions:

- Forms the arc plasma.
- Stabilises the arc roots on the material surface.
- Ensures smooth transfer of molten droplets from the wire to the weld pool.

Thus, the shielding gas will have a substantial effect on the stability of the arc and metal transfer and the behaviour of the weld pool, in particular, its penetration. General purpose shielding gases for MIG welding are mixtures of argon, oxygen and CO_2, and special gas mixtures may contain helium. The gases which are normally used for the various materials are:

- Steels:
  - CO_2,
  - Argon + 2 to 5% oxygen,
  - Argon + 5 to 25% CO_2.
- Non-ferrous:
  - Argon,
  - Argon/helium.

Argon based gases, compared with CO_2, are generally more tolerant to parameter settings and generate lower spatter levels with the dip transfer mode. However, there is a greater risk of lack of fusion defects because these gases are colder. As CO_2 cannot be used in the open arc (pulsed or spray transfer) modes due to high back-plasma forces, argon based gases containing oxygen or CO_2 are normally employed.

Power Supply

Most applications of gas metal arc welding use a constant voltage power supply. As a result, any change in arc length (which is directly related to voltage) results in a large change in heat input and current. A shorter arc length causes a much greater heat input, which makes the wire electrode melt more quickly and thereby restore the original arc length. This helps operators keep the arc length consistent even when manually welding with hand-held welding guns. To achieve a similar effect, sometimes a constant current power source is used in combination with an arc voltage-controlled wire feed unit. In this case, a change in arc length makes the wire feed rate adjust to maintain a relatively constant arc length. In rare circumstances, a constant current power source and a constant wire feed rate unit might be coupled, especially for the welding of metals with high thermal conductivities, such as aluminum. This grants the operator additional control over the heat input into the weld, but requires significant skill to perform successfully.

Alternating current is rarely used with GMAW; instead, direct current is employed and the electrode is generally positively charged. Since the anode tends to have a greater heat concentration, this result in faster melting of the feed wire, which increases weld penetration and welding speed. The polarity can be reversed only when special emissive-coated electrode wires are used, but since these are not popular, a negatively charged electrode is rarely employed.

Electrode

The electrode is a metallic alloy wire, called a MIG wire, whose selection, alloy and size, is based primarily on the composition of the metal being welded, the process variation being used, joint design, and the material surface conditions. Electrode selection greatly influences the mechanical properties of the weld and is a key factor of weld quality. In general the finished weld metal should have mechanical properties similar to those of the base material with no defects such as discontinuities, entrained contaminants or porosity within the weld. To achieve these goals a wide variety of electrodes exist. All commercially available electrodes contain deoxidizing metals such as silicon, manganese, titanium and aluminum in small percentages to help prevent oxygen porosity. Some contain denitriding metals such as titanium and zirconium to avoid nitrogen porosity. Depending on the process variation and base material being welded the diameters of the electrodes used in GMAW typically range from 0.7 to 2.4 mm (0.028 – 0.095 in) but can be as large as 4 mm (0.16 in). The smallest electrodes, generally up to 1.14 mm (0.045 in) are associated with the short-circuiting metal transfer process, while the most common spray-transfer process mode electrodes are usually at least 0.9 mm (0.035 in).

Advantages of Metal Inert Gas Welding (MIG) Welding

- Consumable electrodes are easy to feed.
- No filler rod is needed.
- Welding is simple.
- Inert gas shield protects the weld automatically.

Disadvantages of Metal Inert Gas Welding (MIG) Welding

- Improper welding may lead to the floating of solid impurities over the liquid weld.
- If not handled properly, weld may become porous.
- MIG Welding exposes welders to hazardous gases.
- Care must be taken to avoid the formation of less ductile welds.
- Workpieces and Electrodes should be kept clean before welding.

Submerged Arc Welding

Submerged Arc Welding is one type of welding method where this welding arc can travel under a layer of granular flux. In this type of welding, a tubular electrode otherwise consumable solid can be fed constantly to the weld region. At the same time, a layer of granular fusible flux can be poured over the weld zone which immersed the welding arc as well as defends it from atmospheric pollution.

The granulated flux includes compounds like lime, silica, manganese oxide, calcium fluoride, etc. Whenever the flux is melted, then it turns into conductive as well as offers a current lane among the workpiece & electrode. The solid layer of flux wraps the melted metal totally and stops the sprinkle and covers the strong ultraviolet (UV) radiation vapors generated during the procedure.

Power Supply

DC or AC power source can be used for SAW. And for multiple electrode systems, a combination of DC and AC are commonly used.

The current range utilized in the process is 300 A to 2000 A. for multiple arcs a current of up to 5000 A has also been used.

Main Parts or Equipment

The main parts or equipment used in submerged arc welding (SAW) process are:

- Welding head
- Flux hopper
- Flux
- Electrode wire feed unit
- Electrode
- Flux recovery unit

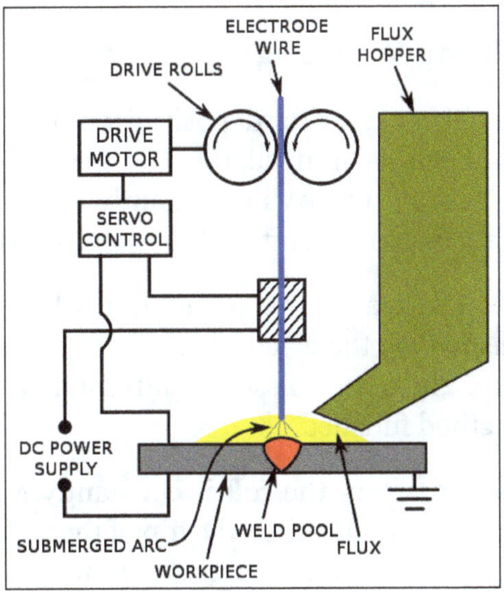

- Welding head: It is used to feed flux and filler metal to the joint to be weld.
- Flux hopper: Here the flux is stored and delivered to the welding joint. It controls the rate of deposition of flux to the welding joint.
- Flux: The weld arc is shielded by granular flux. It consists of lime, silica, manganese oxide, calcium fluoride, oxides of calcium and other compounds. It fed into the weld zone by gravity flow through the nozzle of the welding head.

  When it is molten, it becomes conductive and conducts the current in between the electrode and work. The thick layer of granular flux covers the molten metal completely and prevents the spatter and spark as well as cover up the ultraviolet radiations and fumes which is characteristic of shielded metal arc welding (SMAW) process. The lower part of the flux gets melted and forms a slag on the weld pool. It is removed after the welding process gets completed. The upper part of the flux acts as an insulator and promotes deep penetration of heat to the workpiece.
- Electrode wire feed Unit: It provides a continuous feed of electrode wire to the welding joint. It consists of a reel on which the electrode wire is wounded.
- Electrode: Submerged arc welding uses a consumable electrode which is a coil of bare round wire of diameter 1.5 mm to 10 mm. It is fed automatically through the welding gun. The composition of the electrode depends upon the material being welded. The electrodes are available to weld high carbon steel, mild steel, low and special alloy steels, stainless steel, etc. The electrodes are usually coated with copper to prevent rusting and increases electrical conductivity. They are available in straight length and in coils.
- Flux recovery unit: It collects the unused flux left after welding. And after recovery, it is used again for the welding.

Submerged Arc Welding Working

In this kind of welding, the flux begins for depositing on the joint to be welded. Whenever the flux is cold, then it acts as an insulator. The arc can be started by moving the tool by the work portion. The arc struck will constantly remain below a wide coating of flux, and the generated heat by the arc softens the granular flux.

Once the flux is melted by the heat of the arc, then it will become highly conductive. The flow of current begins to flow the electrode through the melted flux that can be in contact by the atmosphere. The minor dissolved flux alters to wastage slag & which is detached after welding method finished.

At a fixed speed, the electrode from the roll is constantly fed toward the joint to be linked. If linking is partially automatic, then the top of the welding can be moved physically along with the connection. In an automatic submerged arc welding, a separate drive can be used to move the welding top above the stationary job otherwise job moves beneath the head of the stationary welding.

With the help of the self-adjusting arc principle, the length of the arc is kept stable. When the arc length reduces, the arc voltages will increase & this will increase the arc current. Because of this, the rates of burn-off will increase & the arc length will be increased. The reverse phenomenon arises when the arc length rises more than the regular length. For straight penetration as well as for supporting the huge quantity of melted metal a support steel plate otherwise copper may be used.

Advantages

The advantages of Submerged Arc Welding include the following:

- This submerged arc welding process has high (45kg/h) deposit rate.
- In automatic applications.
- Very small welding smoke can be observed.
- No edge training is required.
- This method is used in indoor, and outdoor.
- No chance of weld sprinkles because it is submerged within flux blanket.

Disadvantages

The disadvantages of Submerged Arc Welding include the following:

- The process is incomplete to some particular metals.
- The application is imperfect to direct seams vessels, and pipes.
- The flux usage is hard.
- A health problem can be occurred due to the flux.
- Slag elimination is desirable after welding.

Submerged Arc Welding Applications

The applications of Submerged Arc Welding include the following:

- The Submerged Arc Welding can be used to weld pressure vessels like boilers.
- A lot of structural outlines, pipes, earth moving tools, shipbuilding, railroad construction, and locomotives.
- This type of welding can be used to repair machine parts.

References

- Shielded-metal-arc-welding-basics: fabtechexpo.com, Retrieved 31 March, 2019
- Power-sources-for-shielded-metal-arc-welding-stick, post, welding-articles: longevity-inc.com, Retrieved 14 July, 2019
- Electric-arc-welding-working-principle-types: elprocus.com, Retrieved 17 May, 2019
- Electric-arc-welding: mechanicalstudents.com, Retrieved 19 April, 2019
- Ac-vs-dc-welding: cromweld.com, Retrieved 5 February, 2019
- Gas-tungsten-arc-welding: techminy.com, Retrieved 26 July, 2019
- The-fundamentals-of-gas-tungsten-arc-welding--preparation-consumables-and-equipment-necessary-for-the-process, arcwelding: thefabricator.com, Retrieved 21 May, 2019
- Selecting-a-gtaw-power-source, arcwelding: thefabricator.com, Retrieved 8 January, 2019
- Plasma-arc-welding-principle-working-equipment-types-application-advantages-and-disadvantages: mech4study.com, Retrieved 13 May, 2019
- Electric-arc-welding-working-principle-types: elprocus.com, Retrieved 25 February, 2019

3

Resistance Welding

Resistance welding is the type of welding process that makes use of electrical resistance to produce heat for forming the weld. There are various types of resistance welding such as resistance spot welding, flash butt welding, resistance seam welding, projection welding, etc. This chapter closely examines these different types of resistance welding to provide an extensive understanding of the subject.

Resistance Welding is a liquid state welding method where the metal-to-metal joint can be formed within a liquid state otherwise molten state. This is a thermoelectric method where heat can be generated at the It is a thermo-electric process in which heat is generated at the edge planes of welding plates because of electric resistance and a weld joint can be created by applying low-pressure to these plates. This type of welding uses electric resistance to generate heat. This process is very efficient with pollution free but the applications are limited because of the features like equipment cost is high, and material thickness is limited.

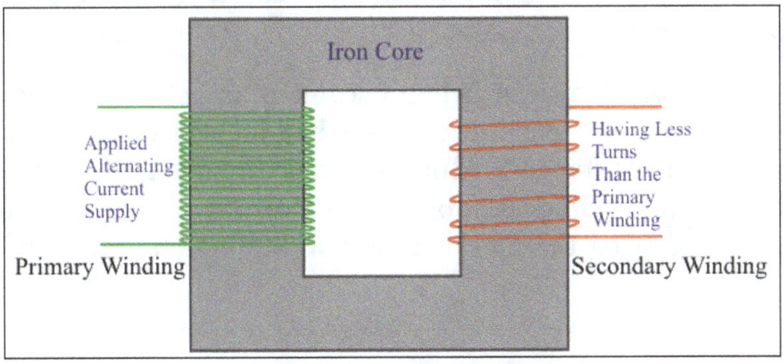

Figure: Step down transformer.

Working of Resistance Welding

The bulk and contact resistance values of the workpieces, electrodes, and their interfaces both cause and affect the amount of heat generated. There are three contact and four bulk resistance values, which, combined, help determine the heat generated. Bulk resistance is a function of temperature. All metals exhibit a Positive Temperature Coefficient (PTC), which means that their bulk resistance increases with temperature. Bulk resistance becomes a factor in longer welds.

Contact resistance is a function of the extent to which two surfaces mate intimately or come in contact. Contact resistance is an important factor in the first few milliseconds of a weld. The surfaces of metals are quite rough if they are examined on an molecular scale. When the metals are forced together with a relatively small amount of force, some of the peaks make contact.

On those peaks where the contact pressure is sufficiently high, the oxide layer breaks, forming a limited number of metal-to-metal bridges. The weld current is distributed over a large area as it passes through the bulk metal. However, as it approaches the interface, the current is forced to flow through these metallic bridges. This "necking down" increases the current density, generating enough heat to cause melting. As the first of these bridges melt and collapse, new peaks come into contact, forming new bridges, and additional current paths. The resistance of the molten metal is higher than that of the new bridges so that the current flow transfers from bridge-to-bridge. This process continues until the entire interface is molten. When the current stops, the electrodes rapidly cool the molten metal, which solidifies, forming a weld.

Parameters in Resistance Welding

The principle of resistance welding is the Joule heating law where the heat Q is generated depending on three basic factors as expressed in the following formula:

$$Q = I^2 Rt$$

where, I is the current passing through the metal combination, R is the resistance of the base metals and the contact interfaces, and t is the duration/time of the current flow.

The principle seems simple. However, when it runs in an actual welding process, there are numerous parameters some researchers had identified more than 100, to influence the results of a resistance welding. In order to have a systematic understanding of the resistance welding technology, we have carried out a lot of experimental tests and summarized the most influential parameters into the following eight types:

Welding Current

The welding current is the most important parameter in resistance welding which determines the heat generation by a power of square as shown in the formula. The size of the weld nugget increases rapidly with increasing welding current, but too high current will result in expulsions and electrode deteriorations. The figure shows the typical types of the welding current applied in resistance welding including the single phase alternating current (AC) that is still the most used in production, the three phase direct current (DC), the condensator discharge (CD), and the newly developed middle frequency inverter DC. Usually the root mean square (RMS) values of the welding current are used in the machine parameter settings and the process controls. It is often the tedious job of the welding engineers to find the optimized welding current and time for each individual welding application.

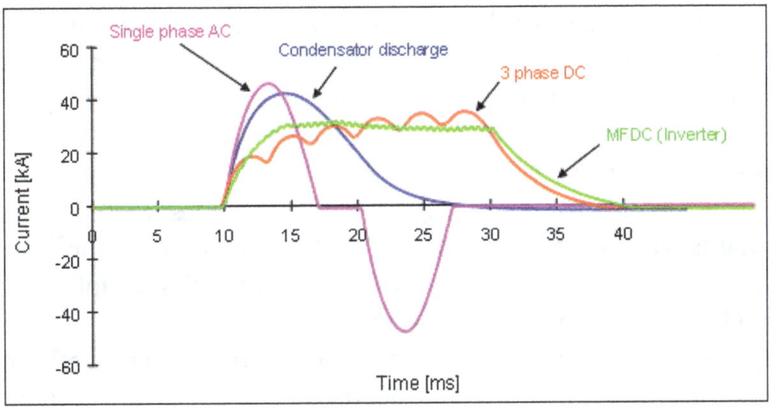

Welding Time

The heat generation is directly proportional to the welding time. Due to the heat transfer from the weld zone to the base metals and to the electrodes, as well as the heat loss from the free surfaces to the surroundings, a minimum welding current as well as a minimum welding time will be needed to make a weld. If the welding current is too low, simply increasing the welding time alone will not produce a weld. When the welding current is high enough, the size of the weld nugget increases with increasing welding time until it reaches a size similar to the electrode tip contact area. If the welding time is prolonged, expulsion will occur or in the worst cases the electrode may stick to the workpiece.

Welding Force

The welding force influences the resistance welding process by its effect on the contact resistance at the interfaces and on the contact area due to deformation of materials. The workpieces must be compressed with a certain force at the weld zone to enable the passage of the current. If the welding force is too low, expulsion may occur immediately after starting the welding current due to fact that the contact resistance is too high, resulting in rapid heat generation. If the welding force is high, the contact area will be large resulting in low current density and low contact resistance that will reduce heat generation and the size of weld nugget. In projection welding, the welding force causes the collapse of the projection in the workpiece, which changes the contact area and thereby the contact resistance and the current density. It further influences the heat development and the welding results.

Contact Resistance

The contact resistance at the weld interface is the most influential parameter related to materials. It however has highly dynamic interaction with the process parameters. The figure shows the measured contact resistance of mild steel at different temperatures and different pressures. It is noticed that the contact resistance generally decreases with increasing temperature but has a local ridge around 300 °C, and it decreases almost proportionally with increasing pressure. All metals have rough surfaces in micro scale. When the welding force increases, the contact pressure increases thereby the real contact area at the interface increases due to deformation of the rough surface asperities. Therefore the contact resistance at the interface decreases which reduces the heat generation and the size of weld nugget. On the metal surfaces, there are also oxides, water vapour, oil, dirt and other contaminants. When the temperature increases, some of the surface contaminants (mainly water and oil based ones) will be burned off in the first couple of cycles, and the metals will also be softened at high temperatures. Thus the contact resistance generally decreases with increasing temperature. Even though the contact resistance has most significant influence only in the first couple of cycles, it has a decisive influence on the heat distribution due to the initial heat generation and distribution.

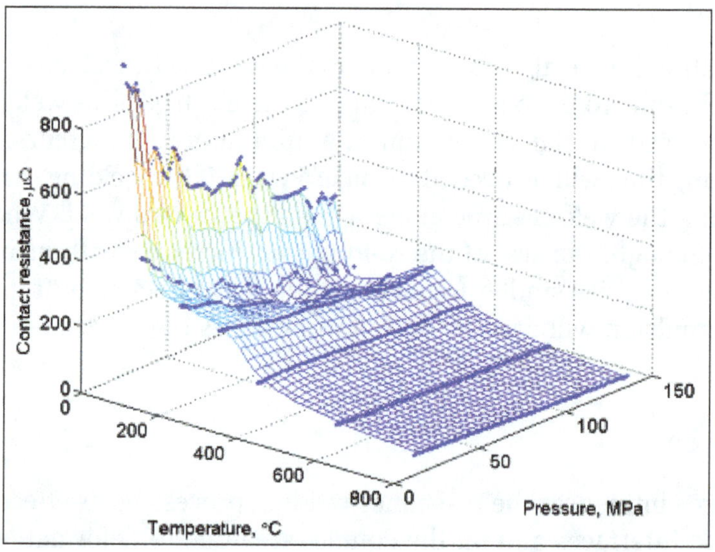

Materials Properties

Nearly all material properties change with temperature which adds to the dynamics of the resistance welding process. The resistivity of material influences the heat generation. The thermal conductivity and the heat capacity influence the heat transfer. In metals such as silver and copper with low resistivity and high thermal conductivity, little heat is generated even with high welding current and also quickly transferred away. They are rather difficult to weld with resistance welding. On the other hand, they can be good materials for electrodes. When dissimilar metals are welded, more heat will be generated in the metal with higher resistivity. This should be considered when designing the weld parts in projection welding and selecting the forms of the electrodes in spot welding. Hardness of material also influences the contact resistance. Harder metals (with higher yield stress) will result in higher contact resistance at the same welding force due to the rough surface asperities being more difficult to deform, resulting in a smaller real contact area. Electrode materials have also been used to influence the heat balance in resistance welding, especially for joining light and non-ferrous metals.

Surface Coatings

Most surface coatings are applied for protection of corrosion or as a substrate for further surface treatment. These surface coatings often complicate the welding process. Special process parameter adjustments have to be made according to individual types of the surface coatings. Some surface coatings are introduced for facilitating the welding of difficult material combinations. These surface coatings are strategically selected to bring the heat balance to the weld interface. Most of the surface coatings will be squeezed out during welding, some will remain at the weld interface as a braze metal.

Geometry and Dimensions

The geometry and dimensions of the electrodes and workpieces are very important, since they influence the current density distribution and thus the results of resistance welding. The geometry of electrodes in spot welding controls the current density and the resulting size of the weld nugget. Different thicknesses of metal sheets need different welding currents and other process parameter settings. The design of the local projection geometry of the workpieces is critical in projection welding, which should be considered together with the material properties especially when joining dissimilar metals. In principle, the embossment or projection should be placed on the material with the lower resistivity in order to get a better heat balance at the weld interface.

Welding Machine Characteristics

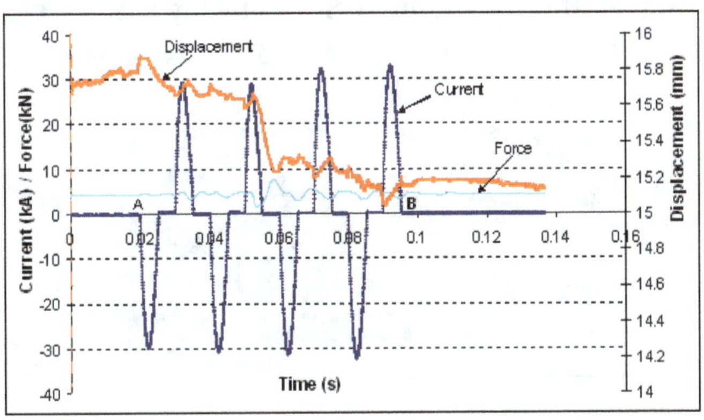

The electrical and mechanical characteristics of the welding machine have a significant influence on resistance welding processes. The electrical characteristics include the dynamic reaction time of welding current and the magnetic/inductive losses due to the size of the welding window and the amount of magnetic materials in the throat. The up-slope time of a welding machine can be very critical in micro resistance welding as the total welding time is often extremely short. The magnetic loss in spot welding is one of the important factors to consider in process controls. The mechanical characteristics include the speed and acceleration of the electrode follow-up as well as the stiffness of the loading frame/arms. If the follow-up of the electrode is too slow, expulsion may easily occur in projection welding. The figure shows measured process parameters in a projection welding process, which include the dynamic curves of the welding current, the welding force and the displacement of the electrode, where the sharp movement corresponds to the collapse of the projection in the workpiece.

Electrode Degradation and Tip Dressing

The resistance welding process is characterized with a high current passing through the materials to be welded between the electrodes under pressure for generating concentrated heat to form a weld. This highly concentrated heat also causes problems to the electrode tips with increasing number of welds.

Mechanisms of Electrode Degradation

The severe conditions of high current and pressure during resistance welding expose the electrode tips at a high risk of degradation. The photo to the right shows a comparison of the new and used electrode tips in spot welding of galvanized steel sheets. With increasing number of welds, there will be two major changes in the electrode tips:

- Geometric changes: The electrode tip diameter will increase due to deformation and wear, such as mushrooming, pitting or local material removal by picking up.
- Metallurgical changes: The material properties near the tip surface will also change during resistance welding such as alloying with sheet and coating materials, and recrystallization and softening by overheating.

New tip Used tip

Effects of Electrode Degradation

The increasing tip diameter will result in larger contact area between electrode and sheet thereby reducing the current density passing through the weld interface. At the same time, alloying of the electrode material with sheet and coating materials at the tip surface will reduce the conductivity of the electrode tip thereby also drag heat concentration away from the weld interface. Both effects lead to progressively reducing weld nugget sizes. After a certain number of welds, the resulted weld nugget will drop to below the minimum nugget size required for the weld quality as shown in the graph to the right. The number of welds achievable until the resulted weld nugget sizes dropping to the limit of weld quality is called the "electrode life". This is dependent on the form and material of electrodes, the materials to be welded, surface coatings, and the interactions of dynamic welding process parameters.

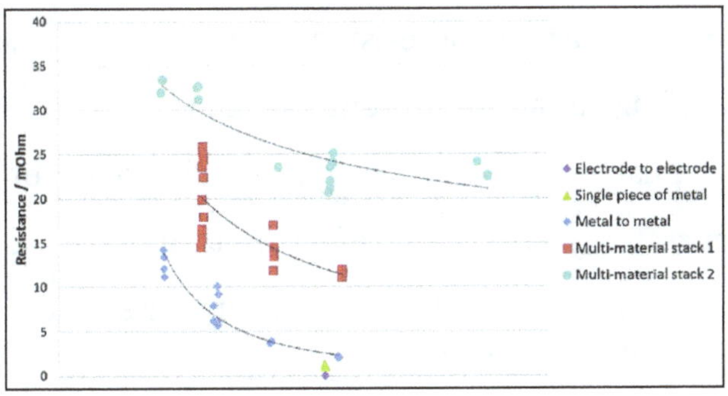

Step Current and Tip-dressing

Apart from adopting new materials and new designs of electrodes, two methods are generally used in production for compensating the electrode degradation in order to maintain the weld quality and increase the electrode life:

- Step current
- Electrode tip-dressing

Step current is a method to plan the spot welding process with a stepwise increasing weld current at each certain number of welds to compensate the loss of current density due to increasing tip diameter as shown in the graph to the right. The higher current needed with the larger tip diameter may be optimized through welding tests or by support of numerical simulations. In this way, more welds can be achieved without replacing electrodes hence a prolonged electrode life.

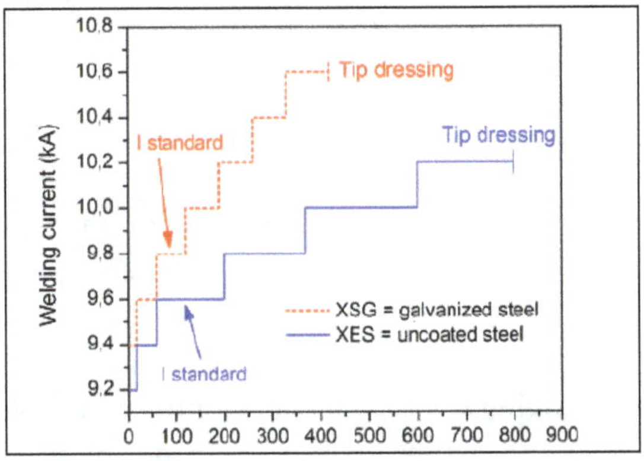

Electrode tip-dressing is a method to mechanically re-shape or abrasively clean the electrode tip after a certain number of welds to preserve nearly the same initial tip diameter and surface conditions. In this way, the welding process can be controlled at the same process parameters or slightly regulated by an adaptive control system to maintain consistent weld quality.

Advantages

Following are the key advantages of of resistance welding:

- Resistance welding has a very short process time.
- It has no consumables, such as brazing materials, solder, or welding rods.
- It provides operator safety because of low voltage.
- It does not create a mess like other welding processes and gives a clean and environmentally friendly production.
- It makes a reliable electro-mechanical joint.

Disadvantages

- The electrodes have to be able to reach both sides of the pieces of metal that are being joined together. A particular spot welding machine will be able to hold only a certain thickness of metal, usually 5 to 50 inches and although the position of the electrodes can be adjusted, there will be only a limited amount of movement in most electrode holders.
- The size and shapes of the electrodes will determine the size and strength of the weld. The join forms only at the spot where the electrodes are in contact with the metal. If the current is not strong enough, hot enough or the metal is not held together with enough force, the spot weld may be small or weak.
- Warping and loss of fatigue strength can occur around the point where metal has been spot welded. The appearance of the joint is often rather ugly, and there can be cracks. The metal may also become less resistant to corrosion.

Application

Spot

Resistance spot welding is made by passing current, for a timed interval, through the workpieces from electrodes which apply the welding force.

Seam

Seam welding consists of making a series of overlapping spot welds by using one or two rotating wheel electrodes without opening the electrodes between spots.

Cross Wire

In practice, it usually consists of welding a number parallel wire at right angles to one or more wires or rods.

Upset

Is a resistance welding process which produces coalescence simultaneously over the entire area of abutting surfaces or progressively along a joint.

Resistance Brazing

Resistance brazing is a method of low-temperature brazing by the application of heat and force to the parts to be joined using heat generated by the current in either the parts themselves, in the dies holding the parts or the electrodes making contact with the area to be brazed, or both.

Resistance Spot Welding

The resistance spot welding process use two copper alloy electrode to concentrate weld current and forced between works to be welded. Produce coalescence in one spot for heat generated from resistance to the electrical current through work piece held together under pressure by electrode. It process usually the thickness range of sheet metal is 0.5 to 3.0 mm. the small spot is quickly heated until melting point, in this reason a nugget of welded metal after power supply removed. When the current is applied too long, which result in molten metal become expelled as welded splash (or) can be made a have right through the material become welded. In this spot welding used for welding steel sheet metal.

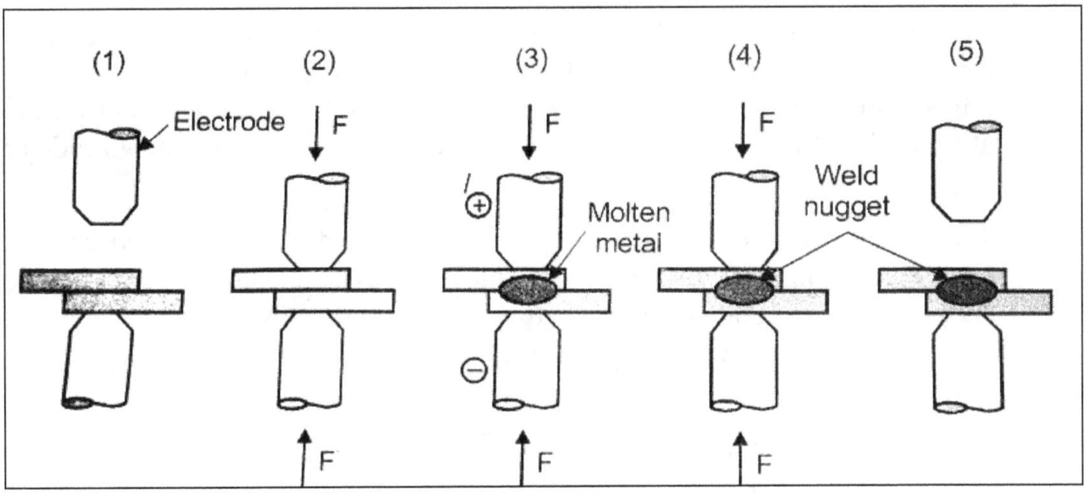

Figure: Resistance spot welding process.

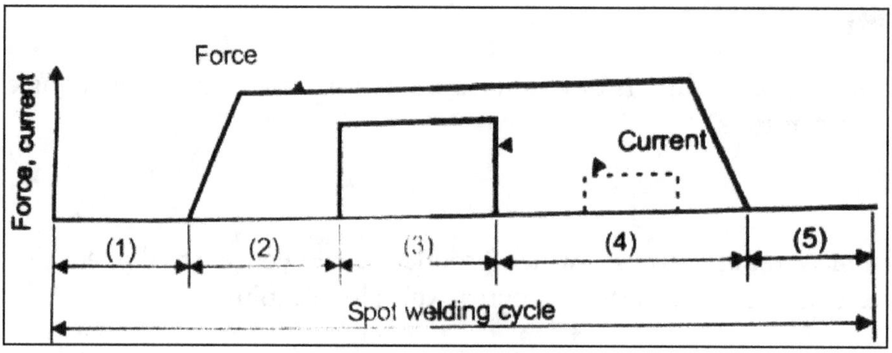

- Parts inserted between open electrodes.
- Electrodes close and force is applied.
- Weld time (current is switched).
- Current is turned off but force is maintained.
- Electrodes are opened, and the welded assembly is removed.

Resistance Spot Welding Step by Step Procedure

- The weld sheet metal is cleaned together and that are free from dirt, oxide, grease and oxides.
- The electrode tip will be cleaned it should be conduct current; otherwise the electric resistance is generated between work piece and electrode.
- Both side of plate, the electrode are stick. When the pressures are created to the electrodes and maintain a particular time period is known as sequence of time for before starting of weld.
- Now, the current is passed through electrode. The weld time is, time of application of time taken. At the weld time during the process pressure to be maintained and weld current range of 3000 A to 100000 A, it passed to electrode at time of 0.1 to 0.5 sec. The step down transformer controls the magnitude of high current possible.
- The current is passed one electrode to another electrode through the work piece, a small area contact between the work pieces. This spot is known as resistance spot. The spot generate the temperature and increase above 900°.
- After weld time completed, the current is shout down and pressure will be maintain a particular brief of time is called as hold time. So the heat metal solidified and form of weld hugest.
- After brief time the pressure released, then work piece is removed.

- Now water is supplied to the electrode because of avoid overheating of electrode. In this welding process, set proper welding current and time before welding.

The size and shape of weld gradually limited by size and contour of electrode. It welding configuration can be relatively simple and inexpensive. Foot operated spot welding machine consist of as shown in figure. That should be having two jaws. One jaw is stationary (it consist of lower) and another jaw is movable (upper jaw). In between the two jaws, the sheet metals are placed and force is applied to per jaw through foot pedal and livers. The motion and pressure are provided from hydraulic or pneumatic operation.

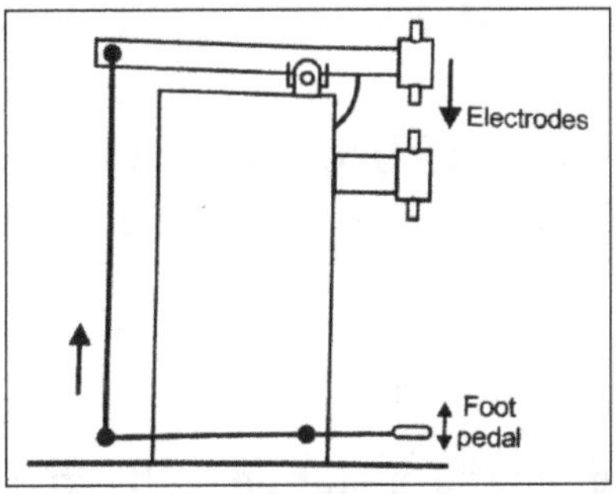

Figure: Foot operated spot welding.

Materials Appropriate for Spot Welding

Due to its lower thermal conductivity and higher electrical resistance, steel is comparatively easy to spot weld, with low carbon steel being most suited to spot welding. However, high carbon content steels (Carbon equivalence > 0.4wt%) are prone to poor fracture toughness or cracking in the welds as they tend to form hard and brittle microstructures.

Galvanised steel (zinc coated) requires slightly higher welding currents to weld than uncoated steels. Also, with zinc alloys, the copper electrodes rapidly degrade the surface and lead to a loss of weld quality. When spot welding zinc coated steels, electrodes must either be frequently exchanged or the electrode tip surface should be 'dressed', where a cutter removes contaminated material to expose a clean copper surface and reshapes the electrode. Other materials commonly spot welded include stainless steels (in particular austenitic and ferritic grades), nickel alloys and titanium.

Although aluminium has a thermal conductivity and electrical resistance close to that of copper, the melting point for aluminium is lower, which means welding is possible. However, due to its low resistance, very high levels of current need to be used when welding aluminium (in the order of two to three times higher than for steel of equivalent thickness).

In addition, aluminium degrades the surface of copper electrodes within a very small number of welds, meaning that stable high quality welding is very hard to achieve. For this reason, only specialist applications of aluminium spot welding are currently found in industry. Various new technology developments are emerging to help enable stable high quality spot welding in aluminium.

Copper and its alloys can also be joined by resistance spot welding, although spot welding copper cannot be easily achieved with conventional copper alloy spot welding electrodes, as heat generation in the electrodes and work piece are very similar. The solution to welding copper is to use an electrode made of an alloy with a high electrical resistance and a melting temperature far in excess of the melting point of copper (much greater than 1080 °C). Electrode materials typically used for spot welding copper include molybdenum and tungsten.

Advantages of Resistance Spot Welding

- Comparatively Low cost.
- Resistance Spot Welding (RSW) method doesn't need highly skilled worker.
- Distortion or warping of parts is eliminated though it leaves some depressions or indentation.
- The joint made is highly uniform.
- Automatic or semi-automatic operation both can be done.
- There is no need for edge preparation.
- Welding can be done in quick succession. It just needs a few seconds to make the joint.

Disadvantages of Resistance Spot Welding

- The equipment cost is high so it can have an effect on the initial cost.
- Skilled welders or technicians are needed for the maintenance and controlling.
- Some metals need special surface preparation for making the RSW a success.
- The thick jobs are not easy to weld.

Applications of Resistance Spot Welding

- Spot welding of thick steel plates has been done and it has replaced the need for riveting.
- The welding of two or more sheet metals can be joined by mechanical means more economically by using the spot welding methods. We don't need gas tight joints.
- Spot welding can be used for attaching braces, pads or clips with cases, bases and covers which are mainly product of sheet metal forming.
- Automobile and aircraft industries rely greatly of spot welding these days.

Types of Spot Welding

Variant Series Welding

When a workpiece is approachable to the electrodes only from one side, as in box-like construction, then it is possible to make spot welds provided a copper plate can be placed against the other side, as shown in figure. Such an arrangement is referred to as series welding.

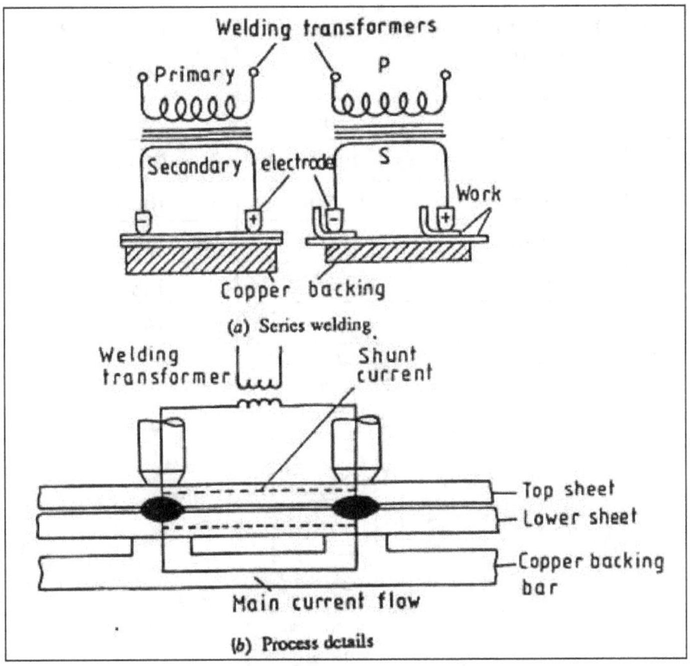

Figure: Series welding setups and process details.

In this process two electrodes are placed 50 to 150 mm apart and force is applied to press the workpiece against the copper backing bar. Ideally the current should flow from one electrode to the other through the copper backing to produce two spot welds; however in practice shunting of current through the workpieces is quite heavy reducing the effective current to a considerably low level as shown in table for welding of two mild steel sheets each 0.9 mm thick under an electrode force of 1.7 kN when a current of 10700 A flows for 12 Hz. Thus the current in the upper sheet is high for the first few cycles because of high interfacial resistance forming part of electrical circuit. However, at the end of the weld time the current is shared equally by the two sheets.

Multiple Spot Welding

Multiple spot welding machines are used for high-volume production work such as the automotive industry. These are press-type machines in which the required number of electrodes are mounted usually with each electrode assembly connected to a separate transformer; however the electrodes normally operate in a sequential order so that all electrodes do not carry current at the same time.

Generally such machines are highly specialised and thus their hydraulic and electrical systems are designed specifically for welding the required range of metal thickness. Figure shows two typical arrangements for multiple spot welding.

Table: Current passing through main and shunt paths in series welding.

Electrode spacing (mm)	Period during weld time	Upper sheet (A)	Lower sheet (A)	Backing (A)
50	Start	3200	2000	5500
	End	2200	2100	6400
100	Start	3200	1600	6800
	End	1750	1700	7250
150	Start	1800	1450	7450
	End	1450	1500	7750

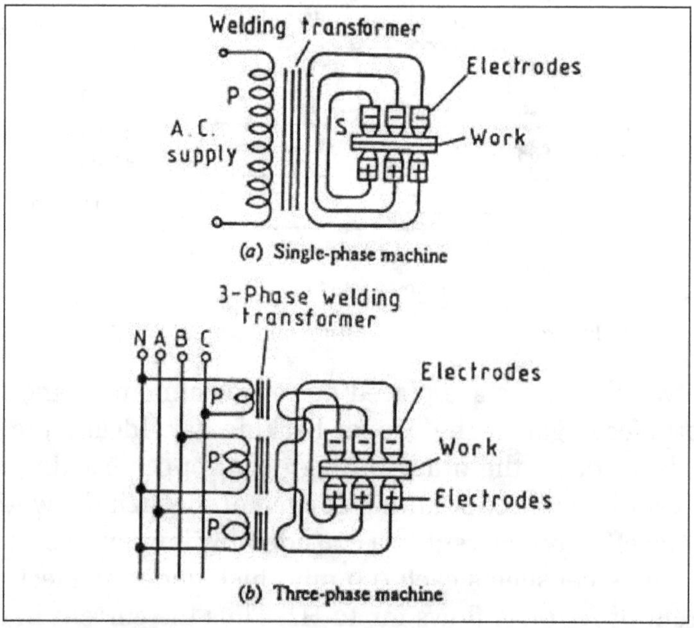

Figure: Single and three phase welding transformers for multiple spot welding.

Pulsation Spot Welding

The main difference, between pulsation spot welding and the conventional spot welding is that instead of a single current flow period in the weld cycle, there is an intermittent flow of current with no current flow intervals in-between as shown in figure. This results in increased electrode life and sounder welds on two pieces of steels thicker than 3 mm and on multiple layer welds.

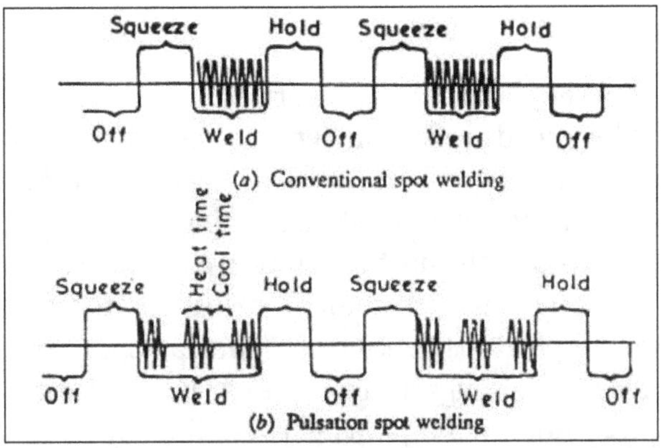

Figure: Nature of weld cycle times in conventional and pulsation spot welding.

Projection Welding

Projection welding process is a one type of resistance welding. In this process produces coalescence of welding metals, the heat generated from electric resistance to current through the work piece hold together with pressure applied to electrodes.

In this projection welding is more or less similar of spot welding process. The two welding work piece are hold with in pressure applied in electrodes. The projection are embossed (top part) in the work piece. Now the electrical current is passed through electrodes, when the projecting works metal to be melted and fuse which is contacted. Which projection coalescence will occur and joint the metal reaction of projection. The welded metal to be done. That should be having multiple work piece can be welded with single operation. The strength of welding joint depends on the nature of projection.

The thin flat sheet metal (24-13 gauge) can be welded by using button type projection. The cone type of projection used for 12 to 5 gauge metal sheets. In case, high thickness of work metal welded by using spherical type projection.

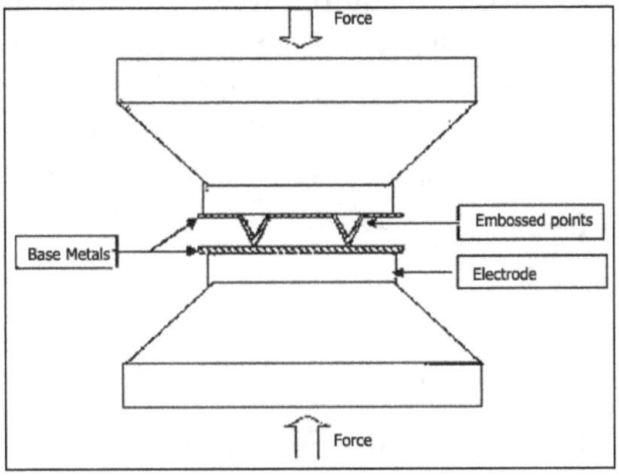

Figure: Projection welding process.

Working Principle

As per the definition, different projections are formed in this welding technique. Here, the metal pieces that are to be joined are kept in between the two electrodes. A larger pressure force is applied to the electrodes. As current is passed through the system, the heat formation takes place due to the internal resistance of the metal workpieces. One point that you must note down here, is that the heat generation takes place due to the internal resistance of the metal workpieces rather than an electric arc. Those projections concentrate the heat. As the pressure applied to the electrodes increases, this projection collapses and the formation of the fused weld nugget takes place. Thus, a quality weld is formed. The exact working of the projection welding can be understood by referring the below image.

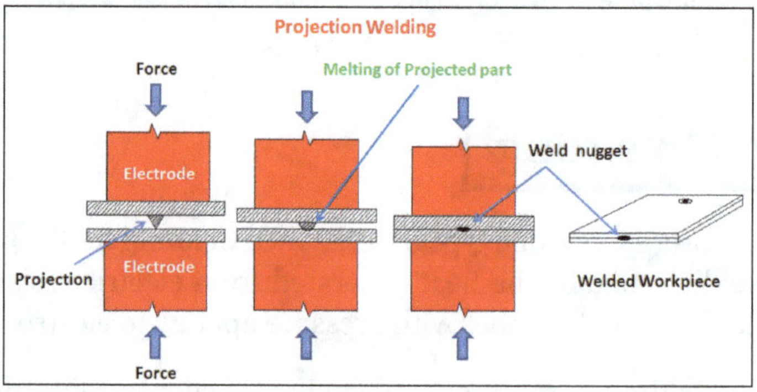

You can see in the above image that the projection means the embossed joints are formed on one of the base metals and then, these base metals are kept in between the two electrodes and force is applied perpendicular to the electrodes.

But as the applied force increases those sharp projections collapses and the formation of the weld takes place at the weld surface. The above image illustrates the formation of the weld nuggets as well as the collapsing of the sharp projections.

In projection welding, the projections of any chape can be used. The only precaution that you must take during the selection of the projections is that they must be able to concentrate the heat to form the required weld. The most used types of projection shapes are button-type projections, cone type projections and spherical type projections. Button type projections are used for the metal sheets having higher thickness. This thickness varies in the range of the 24 to 13 gauge while the cone type projections are used for 12 to 5-gauge metal workpieces. And if you have metal sheets of more thickness spherical type projection is used.

Characteristics

Following are characteristic of the projection welding that you must know:

- A lot of manufacturers prefer the projection welding instead of other welding techniques. So, do you know the exact reason for it? Well, its answer is quite simple. In figure, you can see this welding technique is producing the multiple joints at the same time. So, if you have a very less time and you want the formation of the multiple welds at the same time then, projection welding is a must for you. Therefore, a lot of manufacturers use this welding process instead of others.

- Another important characteristic of the projection welding is that here you get the more electrode life than the spot-welding process. So, it saves your cost of the electrodes and there is no need to change the electrodes in short intervals of time. Some of you might think why the electrode life is longer in the projection welding than the spot welding? In the projection welding, a weld is formed as a result of the heat and the applied force. But in the case of the spot welding, a weld is formed mostly as a result of the heat generation. So, in the projection welding, electrodes are not affected by the current you are supplying. And mostly due to the force applied the required weld formation takes place. It required very less amount of current. And hence, the life of the electrode in the projection welding is much greater than the spot welding.
- Sometimes, plating material is present on the metal surface and this plating may result in the formation of the cracks while welding. Projection welding removes the plating material efficiently and gives a high-quality weld. Those who want to remove the irritating plating material then, go for this welding technique.

Process Control

Once the part is ready for welding at the machine, the three main factors that can be tweaked are the force, power, and time.

Force should be set so it is just enough to make a small indent at the tip of the projection before the weld is fired. Force typically is held constant throughout the weld but can be increased in the latter part of the weld to provide a forging action. Depending on the size and type of materials being welded, time can range from a few milliseconds to 1 second (50/60 cycles). The relationship of power and time is shown in figure. At low power and short time (quadrant III), welds will be weak. High power and longtime (quadrant I) will deliver excessive heating and subsequent meltdown. Most active welding and process tweaking is done in quadrants II and IV.

In quadrant II, the delivery of high power in a short time can lead to arcs, sparks, and electrode sticking. Such arcing can be attributed to poor follow-up of the weld head and constrained movement of the parts being welded. The part in contact with the moving electrode should be completely free to move along with the electrode during weld collapse. In automated manufacturing operations, it is common to have the other end of the part anchored in a housing or fixture. Constrained movement will shrink the size of the process window.

In quadrant IV, the situation is the opposite; the weld head has good follow-up, but the power supply may not be able to deliver power fast enough, leading to a slow collapse of the projection. Longer weld time allows the weld heat to dissipate, it and prevents the weld interface from reaching the desired peak temperature, leading to weak welds.

The total energy (Energy = Power × Time) may be the same for a welding process point in quadrant II and IV, but it is the rate of energy delivery that is critical for projection welding. The process window size can be increased by proper selection of materials (electrodes and parts), part design, welding equipment, and process parameters. A wider process window allows for a robust process that is not sensitive to normal variations in incoming parts.

Other Factors

1. Electrode Life: With projection welding, weld size is not tied to electrode size. Consequently, the electrode size (contact area) can be much bigger than the projection size, resulting in reduced contact pressure and current density at the electrode/part interface. This reduction in pressure and current density helps increase electrode life.

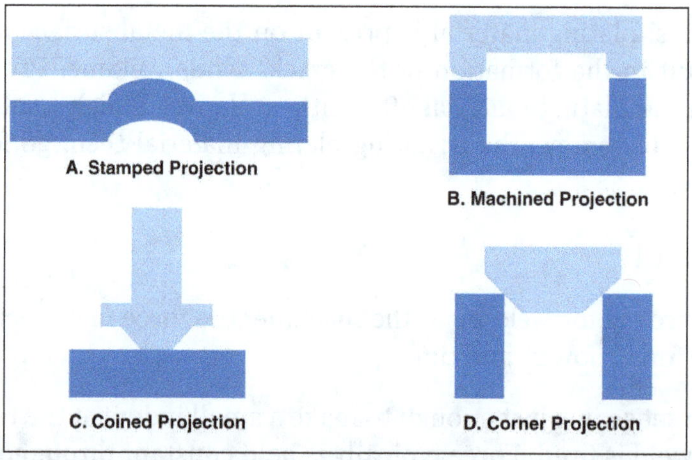

Figure: The four types of projection welds are shown here in schematic form. Machined and corner projections are common in applications for circumferential welds leading to a hermetic joint.

2. Heat Balance: One of the challenges in welding parts that are dissimilar in size or material property is heat balance—the ability to produce equivalent softening and heating on both sides of the weld interface. If the parts being welded are considerably different in size (thermal mass) or material properties (melting point, conductivity), welding with conventional resistance welding can become difficult. In such situations, a projection on the larger or more conductive part helps regain heat balance.

3. Removal of Plating: A key attribute of projection welding is the ability to get rid of plating on the surface and expose parent metal on both sides for a direct bond. Plating material often interferes with welding; for example, zinc plating on the surface of

galvanized steel is not conducive to welding because it has poor bond strength and can actually form cracks in the weld by weakening the grain boundaries.

4. When welded with projection welding, most of the zinc plating gets flashed out during the early part of the weld cycle. The results when welding tin-plated copper parts are similar: Tin plating is squeezed out and allows base copper on both sides to come in contact and form a strong copper-copper, solid-state bond. The plating should have a lower melting point than the base alloy.

5. Multiple Spot Welds: Projection welding is a good option when multiple spot welds have to be made in reasonably close proximity at predetermined locations. It is a straightforward process, and all the welds can be made with a single electrode.

Advantages

- This welding requires a very small supply of current and thus, it saves the electricity usage. So, less electricity requirement and a longer electrode life are the two most prominent benefits of this welding process.
- While doing spot welding there is a limitation on the thickness of the metal that has to be welded. But in this welding, almost metals of all thickness are welded.
- It can be used effectively for welding of the joints which are on the complicate locations.
- The heat balance is an important part of any welding process and this welding gives a good heat balance while welding.

Disadvantages

- This welding process is not applicable for some types of the coppers and brasses.
- Projection formation is a quite complicated process and it takes time to form the projections. It is very difficult to form the spherical projection and a skilled person is required to form such projections. While making those projections, a height of the projection has to be maintained properly.
- This process is not applicable to all types of workpieces. The composition of the metal workpieces has to be considered while this process and it has some limitations.

Application

As Projection welding is mostly used for the mass production. It has many applications such as:

- Automobile industry uses projection welding to a very large extent.

- This welding process is also used for the fan covers and hollow metal doors.
- It is also used for producing the compressor parts and for the semi-conductors.
- Have you heard about the diamond segment welding? In the diamond segment welding, projection welding finds its applications.

Resistance Butt Welding

Resistance butt welding is the simplest form of a group of resistance welding processes that involve the joining of two or more metal parts together in a localised area by the application of heat and pressure. The heat is generated within the material being joined by resistance to the passage of a high current through the metal parts, which are held under a pre-set pressure.

The process is used predominantly to make butt joints in wires and rods up to about 16mm diameter, including small diameter chain. The faces of the pieces to be joined may be flat and parallel or profiled in the case of larger sections. This reduces the initial contact area and further concentrates the heating at the interface. The components are clamped in opposing copper dies, with a small amount of stick-out, and abutted under pressure. Current is passed between the dies causing resistance heating of the weld area. The heat generated during welding depends on the current, the duration of the current, and the resistance. As the resistance is highest at the joint interface, heating is most intense in this area. When the material softens, it deforms under the applied load, giving a solid phase forge weld. No melting occurs. The current is terminated once a pre-set upset length has occurred, or the duration of the current is pre-set. The joint is then allowed to cool slightly under pressure, before the clamps are opened to release the welded component. The weld upset may be left in place or removed, by shearing while still hot or by grinding, depending on the requirements.

Process of Resistance Butt Welding

In resistance butt welding or upset welding the pieces to be welded are held in clamps supported on two platens, one of which is fixed and the other moveable. The ends to be welded touch each other before the current is switched on, as shown in figure. A heavy current is then passed from one piece to another and the contacting faces are heated up due to the contact resistance.

The two pieces are pressed together firmly after the desired welding temperature of 870 to 925 °C is reached. The pressing action which results in the increase in lateral dimension of the workpieces is called upsetting. Upsetting takes place both during and after the current flow. The upsetting action results in welding of end faces with squeezing of a part of the softened metal to form a fin, which can be removed later, if required, by machining.

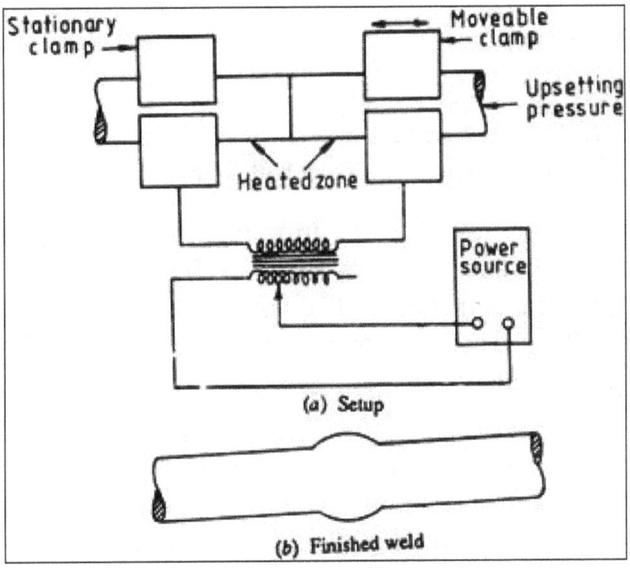

Figure: Upset Welding.

Resistance butt welding is used for end joining of rods, tubes, bars and similar other sections. However, an important application of this process is the large scale production of tubes and pipes at a high rate of production and the process is then referred to as resistance butt-seam welding or simply as electric resistance welding (ERW). In ERW process, the strip for tube making is continuously edge sheared and rolled into tube for forming a longitudinal scam. The continuous welding is carried out as shown in figure.

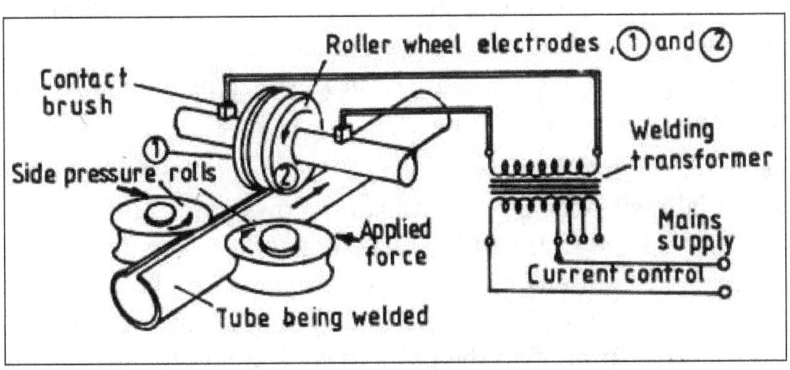

Figure: Electric resistance butt welding process of tube making.

Rings and cylinders can also be welded by electric resistance welding process, as shown in figure. but it is not possible to do so for small sizes because the ring shunts the current heavily. If the circumference is less than 25 times the section thickness there is every possibility of burn through.

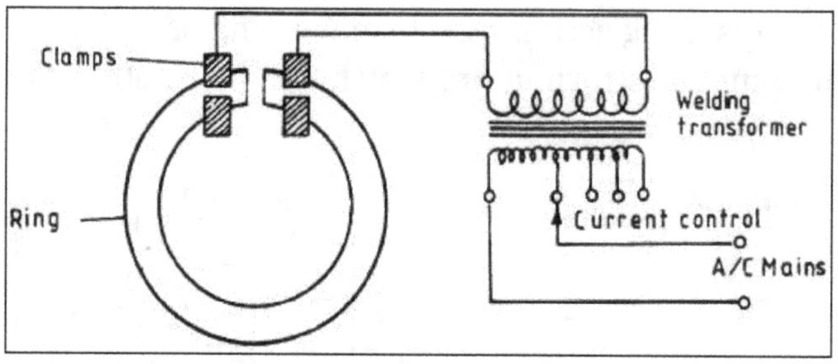

Figure: Electric resistance butt welding of a ring.

Current of upto 40,000 amperes at 5 volt is introduced across the joint by split electrode rollers and the force is applied by the pressure rolls. In this process, both the work motion and current supply are continuous. The flow of current through the shunt path is avoided or reduced by the use of ferrite or wrought iron 'impeder' placed inside the tube.

The maximum speed of production is controlled by the frequency of the current as that decides the number of zero-current periods per unit time. Frequencies upto 350 Hz are commonly used which results in the production rate of about 36m/min. The fin formed due to the extruded metal is continuously removed by cutters and the desired length of the tube or pipe is cut on the production table without any interruption in the process of welding.

Flash Butt Welding

Flash (butt) welding is an electrical resistance welding processes used for joining components, where the energy transfer is provided primarily by the resistance heat from the parts themselves. The components are positioned end-to-end across the full joint area.

A range of section sizes and complex shapes can be joined (for example from bicycle wheel rims to railway rails). The parts to be joined are clamped and brought together slowly while a flashing voltage is applied. The process can be manual, but most welding machines are automatic or semi-automatic.

Where small contacts are made between the components, there is a high current density and the material resistance heats, melts and blows out of the joint in a shower of melted particles, giving the characteristic flashing action. This flashing progressively introduces heating and a softened zone at the component ends, while eliminating oxides and contaminants from the interface. After a pre-set 'burn-off' length of the parent material, a forge force is applied to the parts to consolidate the joint. This produces a forge butt weld with no melted metal remaining in the joint.

Controlling the Flash Welding Process

The key to the flash welding process is the control of the two workpieces toward each other during the flashing operation. Some of the earliest mechanisms for controlling the distance between the two workpieces for the proper flashing action were hand-operated. The operator controlled the current by a thumb button or automatic trip.

The operator became very skilled in controlling the amount of pressure of the two workpieces and resulting flashing action. By observing the heat color of the area and the expulsion of the material, the operator knew when to apply heavier pressure for the second stage (or upset portion) of the flash weld.

Later developments included a motor-driven, variable-speed drive system tied to a gear-reducing unit and cam. This combination provided variable speed, bringing the two workpieces together with acceleration toward the end of the flash welding stage. This acceleration is required because of the increased amount of material expelled as the temperature increases, which in turn increases the resistance between the two materials.

An upset block provided on the cam allowed for the second stage (upset) to be initiated at the proper time. The current was initiated and subsequently turned off by limit switches strategically placed to work in conjunction with the cam, and it could be turned on or off to facilitate the proper heating.

As alloys and special materials were being developed in the industry, they required an entirely different rate of rise on the cam and a change of upset. This caused the cam to be altered or changed when going from one cross section to another or from one alloy to another. For instance, aluminum alloys require a rapid or high-speed upset because of their narrower HAZ. This caused an extremely rapid rise on a cam to accommodate these alloys, and success was mediocre at best. Air over oil cylinders were then used but were limited to smaller cross sections of workpieces.

Another way to control the movement of the two workpieces in flash welding is with a hydraulic/servo valve combination. This provides control and high pressures with rapid acceleration. Flash welding machines with hydraulic operation in use today can generate upset pressures of more than 200 tons.

Most flash welding machines use single-phase AC power supplies. As with the butt weld process, three-phase DC can also be used. Although this does not eliminate the flashing action, it does result in reduced primary current, less material loss, and a narrower HAZ. However, capital costs are higher, and requirements for the process are the same as for DC butt welding.

Developments in Flash Welding

Controls: Flash welding's versatility has increased with the addition of electronic and microprocessor controls that accurately control and monitor the flash welding process. These controls include:

- Feedback information to determine velocity and acceleration of the two workpieces as they come together.

- Current monitoring.
- Flash voltage action.

Preflash and Preheating: Other developments include additional stages before the flash welding. These include a burn-off or preflash that allows the ragged ends of unprepared parts to be squared up before a second stage of preheating. The preheating portion of the flash welding operation allows for heat to be generated into the interface of the weld without a discernable amount of material being lost.

This preheating action is an oscillation of the two workpieces against each other. After the two ends of the workpieces are brought together, the resistance in the material allows heat to be generated, and the two ends are pulled apart (before becoming molten), allowing a cooling effect at the interface of the material.

Once the ends begin to cool and solidify, the process is repeated and continued in a rapid motion until heat is generated back into both workpieces. With this process:

- A large cross-sectional area can be welded with lower current demand.
- Heat can be generated in high-strength alloys without a large loss of material by the flashing action.
- The temperature gradient remains more uniform.

Once the proper HAZ has been reached with the preheat stage, the flash weld stage is then initiated for a short period of time, followed by the upset or forging force.

In many cases, when welding steels with extremely high alloys or carbon content, cracking is possible if the weld is cooled too rapidly to room temperature. If the preheating is insufficient, the cracking may be prevented by a final process in the flash weld called postheating. The postheat cycle can be incorporated in a flash weld machine and adjusted for the desired temperature.

Resistance Seam Welding

Resistance seam welding is a one type of resistance welding process. It is a continuous joint the sheets metal and move through between two rotations electrode wheels. The electric current flowing through the contact between the area, the heats is generated and pressure provided in the wheel to produce a leak tight of weld. For resistance seam welding the roller type of electrode used in the welding process since electrode are consist of wheel. Both the lower and upper electrode wheel is powered. In this process the weld is series of overlapping resistance spot weld made progressively along with a joint rotation the electrode.

The roller wheels rotated over the work piece as welding proceed. The roller electrodes are pressurized and current is flow through bearing of roller electrode. When the electrode wheel can be line with throat of machine.

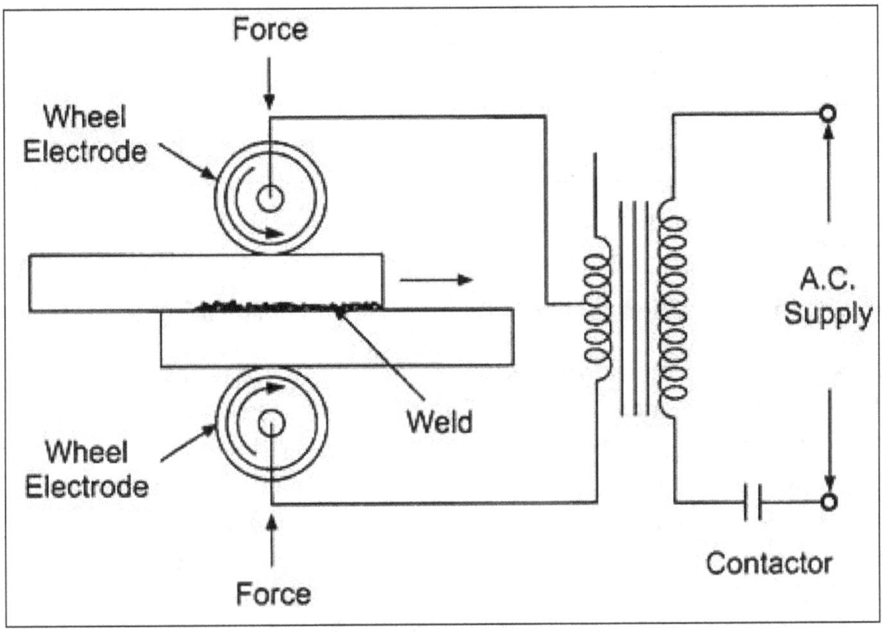

Figure: Lower electrode wheel.

The heat is obtained during the welding process, the heat will be high and rollers are rolled together and cool by water cooling process that reason distortion of roller avoided. The range of pressure in the welding process from 3 MPa to 8.5 MPa based on thickness of work piece. The current range may be as 775 amps/mm^2.

In this operation continuously rather than discretely, the seam along a straight or uniformly curved line form. The operation process sharp corner and similar discontinuities are avoided.

The rather complex system will be required in the resistance seam welding. The travel speed and sequence of current flow are provided for over lapping welds. The welding speed for spot per inch and timing based on pressure, current, speed and size of electrode wheels. It Resistance seam welding process usually providing airtight joint, gasoline tanks, automobile mufflers and some other fabrication of sheet metal container.

Resistance Seam Welding Types

- Conventional resistance seam welding.
- Roll Spot welding.
- Continuous resistance seam welding.

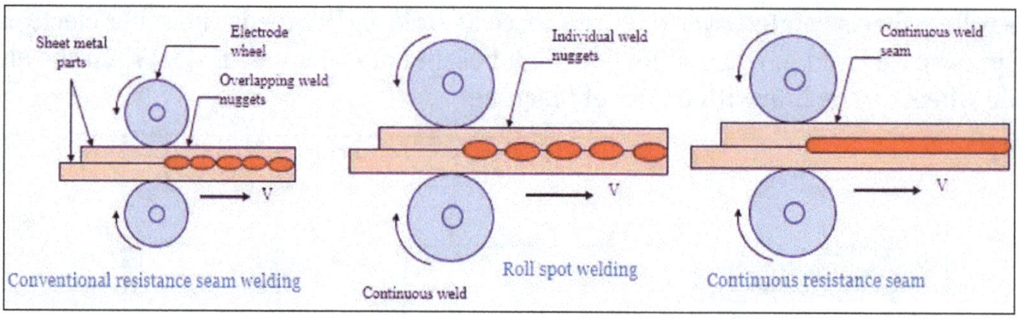

Precaution

A most important precaution that you must take while performing the seam welding:

Whenever we consider any welding process, then there are many precautions that we have to take in order to make that welding process successful. The same case holds true for the seam welding. In case of a seam welding, you have to take a care about the current that you are supplying. If the supplied current is too high then, the interface between the welding plates and roller will be very hot, in such conditions seam welding cannot be done properly. Also, you must be aware of the welding speed while carrying out this process.

To do the seam welding more effectively, sometimes the weld area is washed with water so that the roller type electrodes will remain cool while entire process.

There are two types of seam welding:

- Intermittent Motion Seam welding: Here, the roller moves but its speed is not predetermined as that of the continuous motion seam welding.
- Continuous motion seam welding: In the continuous motion seam welding, the speed of the roller type electrode is always determined first and then, the current is supplied.

Advantages

- This welding is known for forming the clear weld.
- There are only a few types of the welding processes during which no gas formation takes place or not any other fumes are emitted. And the seam welding is one such type of the welding process.
- Another benefit of this welding process is that there is no requirement of the filler material during this welding method.
- Nowadays, every field in the world is becoming automating. Seam welding can be automated so there is no need of addition labors for doing this process.

- Do you want the simultaneous formation of the single seam weld and parallel seam weld? Then, the seam weld is only for you. This welding process produces the parallel seam weld and the single seam weld at the same time.
- With the help of the continuous spot welding, you can form the gas-tight as well as liquid tight joints.

Disadvantages

- This welding process is very difficult to carry out for the sheets of metals which have a thickness larger than 3 mm. So, it is suggested that use this technique if you have sheets of metals which have a thickness smaller than 3 mm. In simple world, it is not applicable if you have metal pieces above a specific thickness.
- You have to follow a particular path in this welding process. That means that rollers always move in a straight line. So, if you want to make a weld at some complex place then, this process is not applicable.
- There is a need of highly-skilled operator or a machine that can handle the speed of the roller as per the situations.
- The machinery that is used in this welding has a very high price. Due to the excessive equipment cost, it is used in a very small proportion.

Applications

- It plays an important role in the manufacturing of all types of the barrels as well as almost every type of the exhaust system.
- Its large number of use is done in case of aircraft tanks, refrigerators, oil transformers, etc.
- It is also used in the welding of the stainless steel, nickel alloys, and magnesium alloys.

References

- Resistance-welding-working-principle-types-and-applications: elprocus.com, Retrieved 16 January, 2019
- Resistance-welding: theweldingmaster.com, Retrieved 29 March, 2019
- Resistance-welding, technology: swantec.com, Retrieved 29 June, 2019
- Resistance-spot-welding: techminy.com, Retrieved 30 April, 2019
- What-is-spot-welding, faqs, technical-knowledge: twi-global.com, Retrieved 14 July, 2019
- Resistance-spot-welding-rsw-working-principle: blogspot.com, Retrieved 11 January, 2019
- What-is-projection-welding-working-principle-advantages-disadvantages-and-application: theweldingmaster.com, Retrieved 18 April, 2019

Solid State Welding

The welding process that joins the pieces of metals with the use of high pressure and a temperature below the melting point of the parent metal is defined as solid state welding. It includes explosive welding, forge welding, friction welding, solid state diffusion welding, etc. This chapter discusses in detail the concepts related to solid state welding.

Solid state welding processes are those welding processes in which no external heat is applied and the joining of work piece takes place in solid state. In these welding processes, no filler metal is used and welding does not involve molten state of base material or filler material. The weld formation take place due to inter molecular diffusion process in which, the interface molecules of work pieces flows from high concentration region to low concentration region due to applied pressure. Some methods are used to produce heat which accelerates diffusion process at mating surfaces. This type of welding processes does not affect the mechanical or physical properties of parent material thus these are widely used in industries application. These are ideal joining processes for heat sensitive material.

Solid State Welding Process

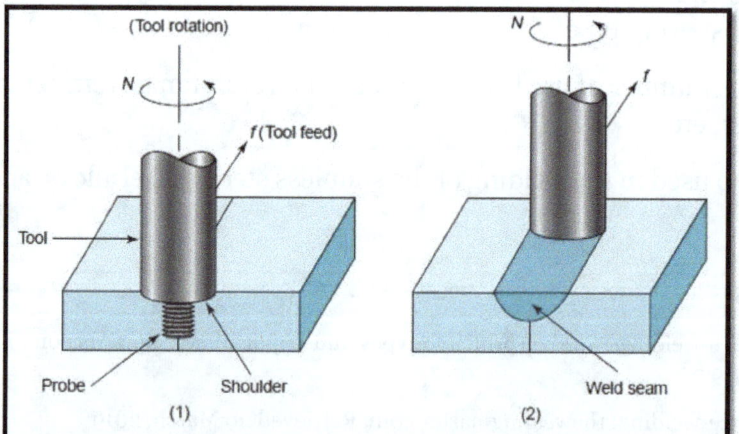

(1) Rotating tool just prior to feeding into joint (2) partially completed weld seam

Principle

Mainly metal joining process is classified into two types. The first one is liquid state welding process in which, metal work pieces are heated until its melting temperature and the metal to metal joint form due to solidification of melted work pieces together. Sometimes this process uses filler material which is melted with base material and allow to solidity, which make a permanent strong joint. The other one is solid state welding process in which no molten or liquid state involves and metal joining process takes place due to application of high pressure into solid state. The metal to metal joint form due to inter molecular diffusion process at the interface surfaces.

Application

Solid state weldings are used in everywhere in mechanical industries. It is widely used in aerospace and marine industries for structure work. It is widely used in automobile industries for most of fabrication work. Explosive welding used to join large welding plate which can't be possible or uneconomical by other methods like arc welding or gas welding. Friction welding is used to join tubes, shaft etc. These processes are further used to join dissimilar material used in various industries. Hydraulic piston, connecting rod, drive line, truck roller bushes, pump shaft, gear lever, drill bit etc. are friction welded.

Advantages

- Solid state welding can be easily automated.
- This produces high strength joint without applying external heat.
- They are used to weld both similar and dissimilar material.
- Provide good surface finish.
- They do not use any filler metal or flux as used in arc welding.
- Mostly these processes do not affect properties of parent materials.

Disadvantages

- High equipment cost or set up cost.
- Welding preparation is more critical.
- Complicated and special fixtures are required for different process.
- Mostly, this process can't be used for mass production due to slow welding speed.

Forge Welding

Forge welding is the type of solid state welding process in which two metal pieces are first heated and then, they are joined. During this welding technique, a process of hammering is also being carried out. It is considered as the oldest method of welding. In the previous century, this welding process is done mostly manually. But as the technology is evolving day by day, the automatic forge welding has been invented.

Principle

Forge welding is a solid state welding process in which both the plates are heated quite below its melting temperature. This heating deforms the work pieces plastically. Now a repeated hammering or high pressurize load is applied on these plates together. Due to this high pressure and temperature, inter-molecular diffusion takes place at the interface surface of the plates which make a strong weld joint. This is basic principle of forge welding. One of the basic requirements of this types of welding, is clean interface surface which should be free from oxide or other contaminant particles. To prevent the welding surface from oxidation, flux is used which mixes with the oxide and lower down its melting temperature and viscosity. This allow to flow out the oxide layer during heating and hammering process.

Working

Forge welding was one of the most applied welding methods in ancient time. This is a fundamental welding process of all solid state welding. Its working can be summarized as follow:

- First both the work plates heated together. The heating temperature is about 50 to 90% of its melting temperature. Both the plates are coated with flux.
- Now manual hammering is done by a blacksmith hammer for making a joint. This process is repeated until a proper joint is created.
- For welding large work pieces, mechanical hammering is used which is either driven by electric motor or by using hydraulic mean. Sometime dies are used which provides finished surface.

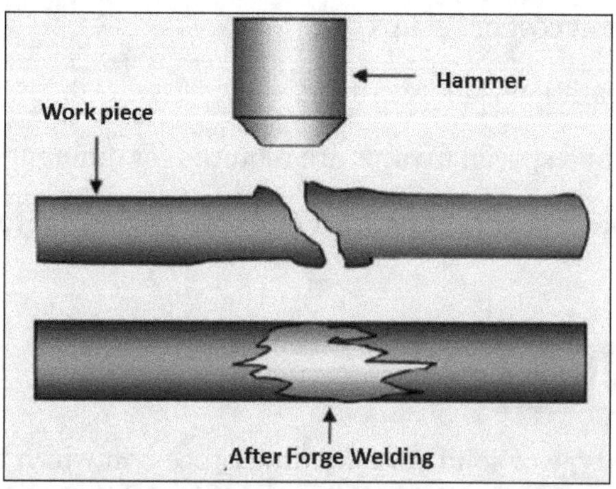

Equipment's used in Forge Welding

Forge or Open Hearth

A typical smith's forge or hearth is shown in figure. This open furnace is used to heat the metal to be shaped to the desired design. This furnace is used for small piece jobs to be forged by hand. Firing may be used for the purpose by Gas, Oil, or Coal. The required air for the fire is supplied under pressure, by a blower (motor driven) through the tuyere in the hearth.

The amount of air supply is controlled by valves near the forge. In the forge, the hood collects the gases of combustion and sends them through a chimney. The air blast to the fire is conducted through the tuyere to the fuel, while the tuyere itself is kept at a lower temperature by the water jacket.

The water circulates by convection. The fire is lit using paper, and sticks of wood. When these are well-alight the fuel is fed into them and a slight air blast is introduced. More fuel is added and the air blast is increased until the fire is suitable for the work. The temperature of the fire can reach 1,500-1,800 °C. Welding temperatures for iron and steel can be reached at about 1,500 °C.

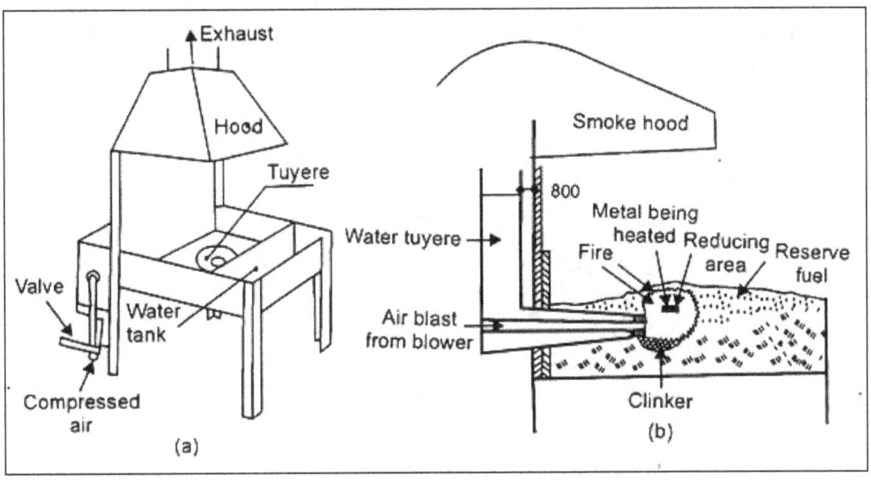

Figure: Forge-hearth or Furnace.

Anvil

An anvil provides the necessary support during forging operations in the different sections of the metal. It is very useful for operation such as bending, swagging, etc. It is made of cast steel, wrought iron or mild steel, with a hardened top layer about 20-30 mm thick.

Some fabricated anvils are used in industries. Such an anvil is made from two pieces of flat- bottomed railway lines about 600 mm long, welded, and cut and shaped together. The Figure shows an anvil with various parts marked on it. Beak is used for bending metal to round shapes. Square hard die hole is used to hold square shank tools like hardies and swages. The punching hole is used for bending small rods and punching holes in the work. Approximate weight of anvil is 75-250 kg.

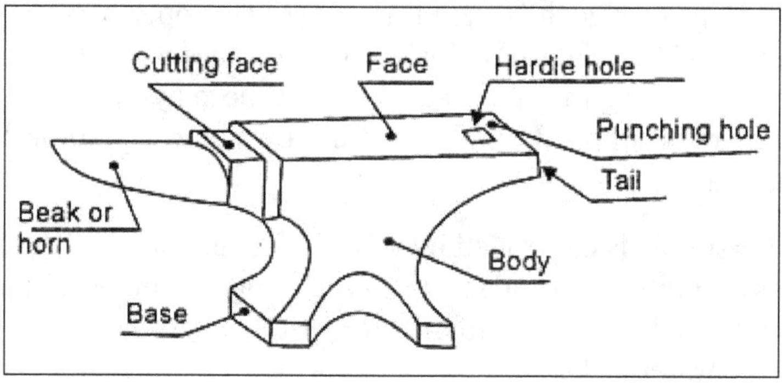

Figure: Anvil.

Swage Block

The swage is an essential equipment of the forging shop. It is used as a support while forming or swaging different shapes of metals and in punching holes. It has a number of slots of different shapes and sizes along its four side-faces and through holes of different shapes and sizes, running from its top to bottom faces. Swage blocks are generally made of Cast Iron or Cast Steel.

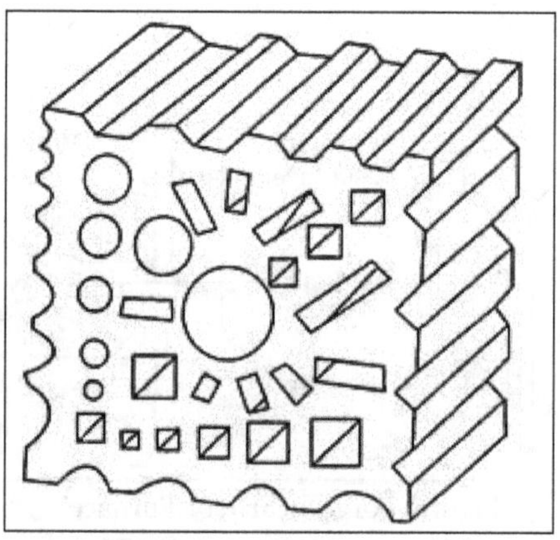

Figure: Swage Block.

Leg Vice

The vice is a device to hold the work or 'Job'. Pressure is applied to the hand of the 'vice' to tighten the work against the fixed jaw. The jaws are made of hardened steel. Leg vice is a heavy-duty vice, and is fixed to the work bench at one end of a leg or set in a concrete base or pillar. It is mainly used for light forging and bending work.

The vice is made of cast iron, strong in compression; but it fractures under shocks and, therefore, should never be hammered.

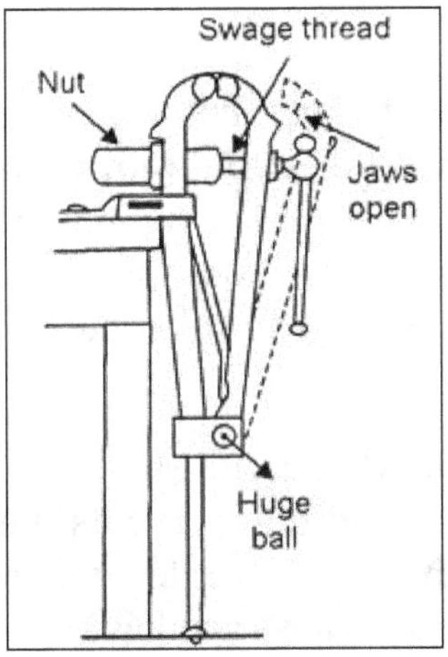

Figure: Leg Vice.

Blacksmithy or Sledge Hammer

Hammers are specified by their weight and different shapes and sizes. The ball pen hammer is used for forging different weights. The sledge hammer is used for heavy work. It has flat ends on either side and weighs 3-8 kg. The length of the handle of a hammer increases with its weight. Figure shows a sledge hammer of forging shop.

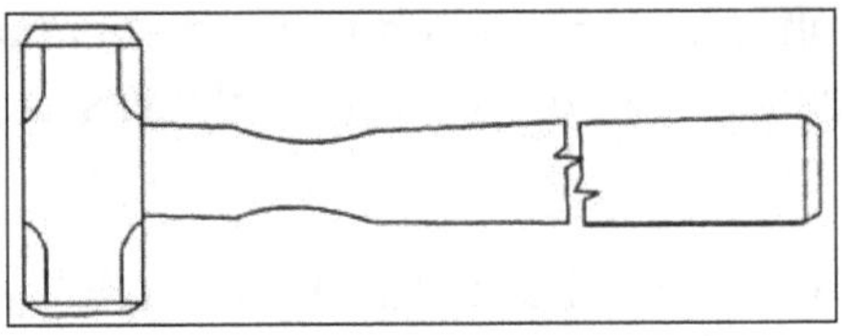

Figure: Sledge Hammer.

Tongs

Tongs are essential for a forging shop. Use proper tongs for proper work. The metal to be forged must be held securely while it is being shaped. A pair of tongs of suitable size and shape must be used for each purpose. Some commonly used tongs in blacksmithy shop are shown in figure.

Tongs are made of M/S (mild steel). Their sizes vary from 40 cm to 80 cm in length, and 0.6 cm to 5.5 cm opening. Flat bit tongs, round bit tongs, square bit and pick up tongs are used in the forging shop.

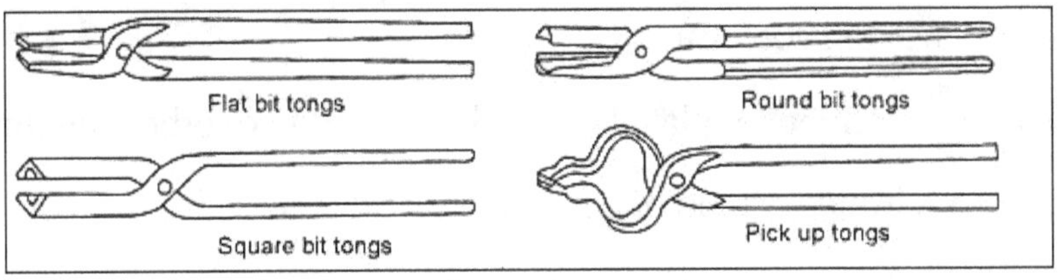

Figure: Tongs.

6 Steps to Forge Weld

- It is of utmost importance to note that welding heat varies, based on the type of steel being used. However, what you want to watch for is a yellow color. Once you get it to that color, you are at the right temperature, if you keep going after this point, your metal can turn into a sparkler. This is when the carbon in the metal begins to oxidize, which will ruin your metal. So first bring your steel to good orange heat.
- Then sprinkle with a flux (like a 20 Mule Team Borax). The flux serves as a low-temperature glassy shield that prevents the oxygen from the steel surface which causes scales on the surface. Scales will prevent your metal from welding. It is advisable to use a flux because you have to a good amount of skill and experience or an oxygen-free burner to weld without scaling your metals.
- After you have fluxed, place the steel back in the heart of the fire, while avoiding direct air blast as you do not want to oxidize your metal.
- Allow the steel get to a lemon yellow color (important: do not look into the fire without using a pair of didymium glasses). It will appear shimmery and almost slippery.
- The pieces of steel should be removed from the heat now as this temperature is just below the spark causing temperature. Sparks mean that your steel is contaminated.
- Now, secure your pieces together with a bit of metal wire until the primary weld forms. After this step, simply use a power hammer or hydraulic press to hammer from one end to the other. You have to be careful not to hit the metal too hard; the hammer blows should be firm and solid but not the same force you would use if you were changing bar shapes. It really depends on the thickness of your pieces. Thicker ones would require more force. Also, you need to be careful not to miss any spots.

Application

- It is used to join steel or iron.
- It is used to manufacture gates, prison cells etc.
- It is widely used in cookware.
- It was used to join boiler plates before introduction of other welding process.
- It was used to weld weapon like sword etc.
- Used to weld shotgun barrels.

Advantages

- It is simple and easy.
- It does not require any costly equipment for weld small pieces.
- It can weld both similar and dissimilar metals.
- Properties of weld joint are similar to base material.
- No filler material required.

Disadvantages

- Only small objects can be weld. Larger objects required large press and heating furnaces, which are not economical.
- High skill required because excessive hammering can damage the welding plates.
- High Welding defects involve.
- It cannot use as mass production.
- Mostly suitable for iron and steel.
- It is a slow welding process.

Friction Welding

In friction welding one piece is rotated and the other is made to rub against it under an axial load resulting in increased friction, heat generation and joining when the pieces are brought to rest under sustained or enhanced axial load, as depicted in figure.

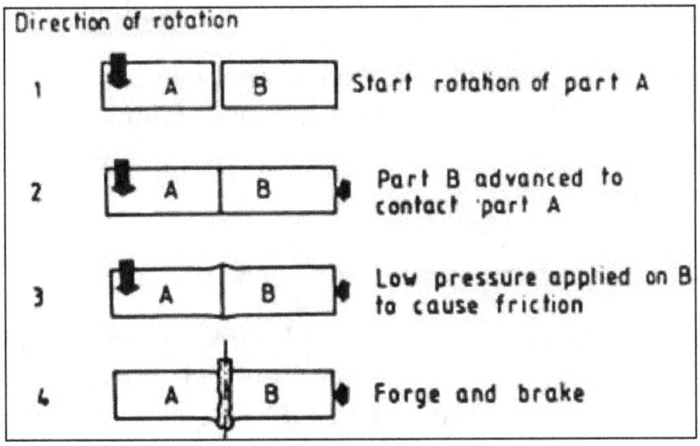

Figure: Sequence of friction welding operation.

Filler metal, flux, or shielding gas is not required in friction welding and the joint is similar in appearance to the ones produced by electrical resistance butt welding processes of flash and upset welding. Normally cylindrical pieces like rods and tubes are welded by this process but its application can be extended to situations wherein one of the components is symmetrical, and can be rotated conveniently. Different modes of the process, as it is used presently, are shown in figure.

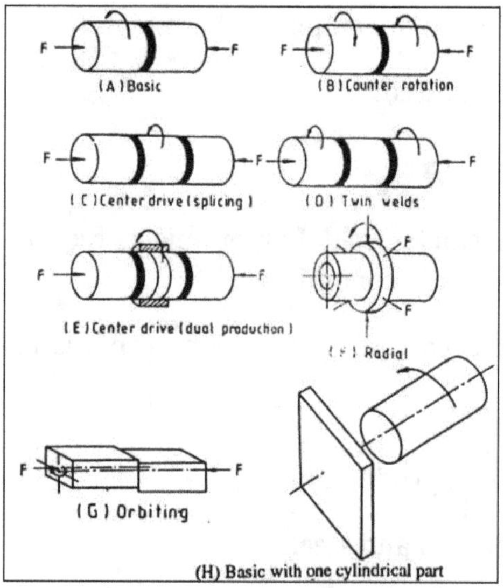

Figure: Different modes of friction welding.

Method A is the simplest and is applicable to most of the steels in the temperature range of 900 – 1300 °C. Method B is used when high relative velocities are required to weld small diameter workpieces. Method C is used for twin welds between two long workpieces which are difficult to rotate. Method F shows what is known as radial welding in which the applied force is perpendicular to the axis of rotation.

The outer ring or sleeve is compressed as it is heated and the pipe wall is supported by an internally expanding mandrel which prevents the penetration of upset metal into the bore of the pipe. This method can also be employed for welding collars to solid shafts.

Method H can be used for welding cylindrical components to plates, for example a rod to a base plate. Method G shows the friction welding of non- circular components; in such a case the pieces are quickly aligned after the motion ceases so that joining is affected when the two edges are still in the plastic state.

Theoretical Considerations for Friction Welding

Chudikov and Vill from Russia are credited with the successful application of friction welding to metals. The fundamental consideration of the process is of course, based on the well-known law that friction force F, is proportional to the applied normal load, L.

Thus,

$$F = \mu L.$$

Where μ is the coefficient of friction which increases with increase in load and is also dependent upon the speed. According to Vill frictional force can be expressed by the following equation,

$$F - \alpha A + \beta L$$

where, A is the area of contact and α and β are constants. For high values of pressure the first term is very small and thus F = βL where β is nearly equal to μ so that the basic law of friction hold well.

From the start of the operation till the weld is completed after applying brakes, the frictional force varies. To study its effects on the different phases of the process it is convenient to do so by analysing the time-torque relationship shown in figure. The initial peak in the torque curve is due to dry friction but soon after follows the second phase of the process wherein seizure and rupture takes place at the high points of the contact.

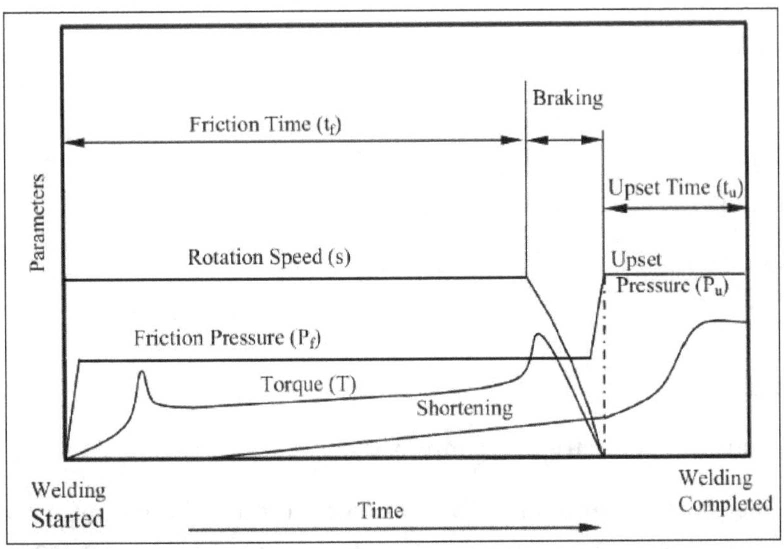

Figure: Time-torque for continuous drive friction welding process.

The average temperature during the second phase is only 100 – 200 °C. The quick rise and the fluctuating nature of the curve are due to the change from marginal or boundary layer friction with μ. = 0.1 to 0.2 to pure friction with μ > 0.3.

The torque in US points of seizure increase and finally molten metal may appear at these points of contact and act as a lubricant and the average temperature of the interface may rise to 900 – 1100 °C.

Only 13% of the total heat is produced during the first two stages ($T_1 + T_2$) while the rest is produced during the third stage (T_3). Increased speed instead of reducing the duration of the process increases it as is evident from figure. This is because enhanced speed results in decreased heating intensity.

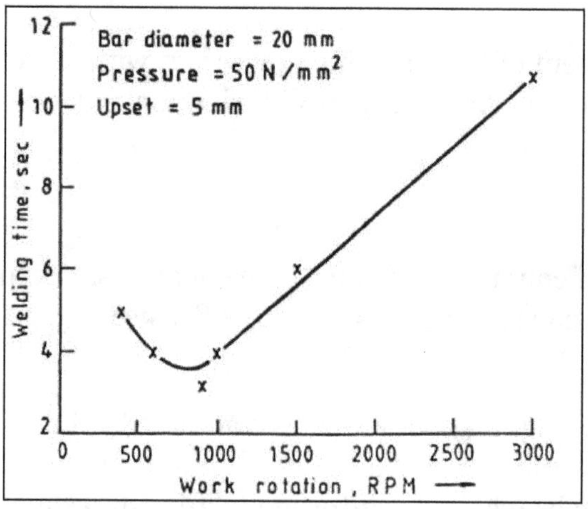

Figure: Effect of rotational speed on duration of welding.

Heat energy generated per unit area of the faying surfaces is given by the following expression:

$$H \approx 2\ PK/nR\ 10^2\ \text{watts/mm}^2$$

where,

H = heat generated, watts/mm^2

P = applied pressure, N/mm^2

R = work radius, mm

n = rpm

K = a constant = 8×10^7 mm^2/min^2 for low carbon steel.

The second stage may cover nearly 30 – 70% of the total time; however this phase is non-productive so it is intended to reduce its time period to increase productivity. This is usually done by increasing the clamping pressure to a maximum possible.

The power required is reported to be proportional to the axial pressure and the duration of the third stage is inversely proportional to the axial pressure. Thus, for optimal results the axial load should be kept low during the initial phase and be increased gradually or may be applied in two stages.

The maximum temperature attained is controlled by the applied axial load as the metal below a specific strength will be squeezed out under a specific load. When viscosity or strength of the plastic metal is low the metal gets thrown out by the centrifugal force under a low axial load for example, in welding of copper.

In friction welding dissimilar metal combinations such as stainless steel to carbon steel, the plane of maximum temperature can move away from the interface; with high speed it moves into the stainless steel so that one-half of the flash is bimetallic. In this case decreasing the rotational speed provides the desired result and at a certain speed the interface again becomes the plane of maximum temperature and thus a plane of maximum shear strain rate.

The applied pressure is perhaps the most important single factor as it controls the temperature and determines the torque required. The heat input rate is proportional to the product of torque and the rotational speed. Rotational speed has to be such that certain minimum or threshold power is exceeded. If the power is above the threshold the process is self-regulating.

If too much power is applied, the width of the sheared zone of metal increases. If the power applied is just above the threshold it will take a long time for the required temperature to be reached and the heat affected zone will be wide. In the table the most important variable is the unit pressure applied during rotation and the recommended values for some of the metals are given.

Table: Friction pressure for different metals.

Metal	Friction pressure N/mm^2
Mild steel	5500
Medium carbon steel	7000
Nickel – chromium steel	11000
Aluminium alloys	8000
Nimonic alloys (80/20 Ni/ Cr)	14000

The sliding velocity varies from zero at the centre of the workpiece to a maximum at the peripheral surface, and the radius at ⅔rd the diameter of the workpiece is used for the calculations. Longer heating times result in more material for forging and for optimal results there should be adequate heated material available for forging when the rotation is stopped. Excessive initial pressure results in excessive squeezing of heated metal leaving only relatively cold metal to be forged when the forging pressure is applied.

Process Characteristics of Friction Welding

One of the two workpieces is rotated at a constant speed throughout the operation except when the brakes are applied at the final stage; the process is therefore often referred to as continuous drive friction welding.

The workpieces rub together under pressure for a predetermined heating time or until a pre-set axial shortening takes place. The drive is then disengaged and the rotation of work is stopped by applying brakes. The axial pressure is maintained or increased, to forge the metal, till the weld cools. The figure shows how the process parameters change during welding when the force at the end is increased to forge the joint. Welds in mild steel can be made by merely maintaining the pressure constant.

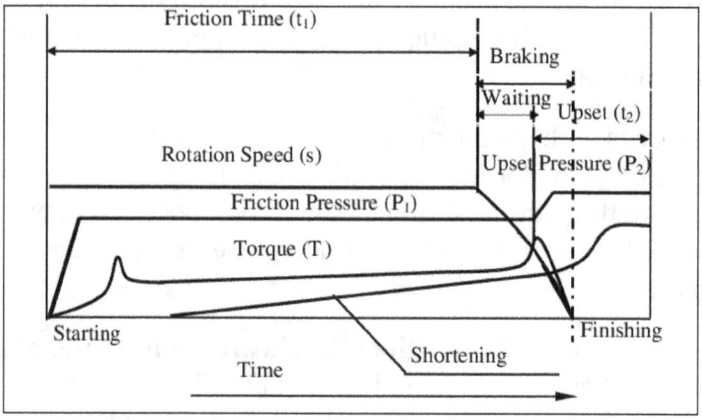

Figure: Variations of process parameters with time for drive friction welding.

With the decrease in speed the thickness of the very hot plasticized band increases and the torque drops to zero as the rotation stops.

Bonding mechanism in friction welding of dissimilar metals is more complex. Due to mechanical mixing and diffusion some alloying possibly occurs in a very narrow band at the interface. The properties of this narrow band may have considerable influence on the overall joint performance. Mechanical mixing and interlocking may also help in bonding. Due to these complexities prediction of weldability of dissimilar metals is very difficult and it needs to be established for a particular application by a series of tests designed specifically for the purpose.

Machines and Equipment Required for Friction Welding

The major components of a friction welding machine, as shown in figure.

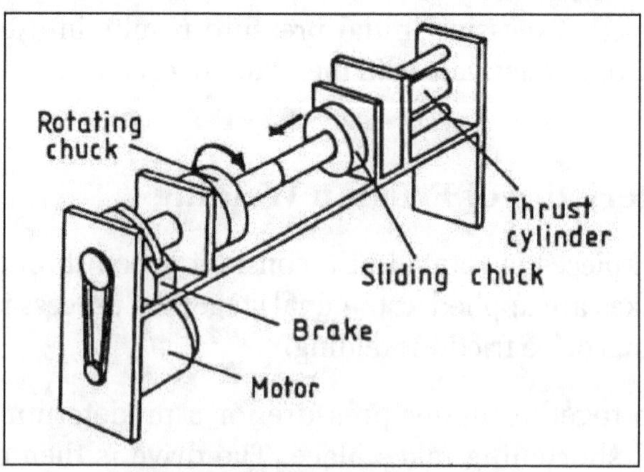

Figure: A friction welding machine.

Include:

- Driven head.
- Clamping arrangements.
- Rotating and upsetting mechanisms.
- Controls.
- Braking mechanism.

One of the workpieces to be welded is held firmly in self-centering head and the other is held in a centering chuck that is mounted on a rotatable spindle which is driven by a motor usually through a variable speed drive.

Rotating chucks must be well balanced, have high strength and provide good gripping power. Collet chucks meet these requirements well and are therefore most frequently used. The gripping mechanism of the chucks must be rigid and resist the applied thrust Serrated gripping jaws are recommended for maximum clamping reliability.

Attempts to use lathe for friction welding were not much successful because it lacked the rigidity of the structure and efficient gripping. A lathe is not designed for endurance need for the friction welding nor is it provided with quick disengagements of the process. Also, to case the braking problem a low moment of inertia is required in the rotating parts.

Friction welding machines are required to control accurately three variables viz., axial thrust, speed of rotation, and degree of upset Plain carbon and low alloy steels require a forge pressure of 15 – 30 N/mm^2, while tungsten steels require pressure in the range of 225 – 400 N/mm^2. These latter values are comparable to the pressure used in flash butt welding. When the rate of application of pressure, by a hydraulic system is found to be low, it is replaced by a pneumatic system.

The duration of welding on a 25 mm diameter bar should be from 5 to 7 seconds. This can be achieved by a surface speed of 75 – 600 m/min which corresponds to about 1000 rpm. Higher speeds may give better impact strength and are, therefore, desirable for hollow sections and high hot strength alloys.

Control of the welding operation may be by time or the extent of upset. The latter method is applied by limit switches arranged to increase the pressure to cause upset after a certain amount of shortening has occurred. Long times tend to allow the heat to spread back to regions behind the interface and therefore result in heavy upsets which are costly to remove after the operation is over. The accuracy of upset is expected to be within 0.1 mm.

Controls by time sequence are found satisfactory for cases where constant surface condition cannot be assured and welds are of secondary importance. When time control is used, high rotational speeds are preferred.

The speed for mild steel is selected on the basis of stock diameter and is given by the expression:

$$Nd = (1.2 \text{ to } 6.0)10^4$$

where, n is the rpm and d is the stock diameter in mm.

The smaller values of the constant refer to welding with high rates of input and in such cases the upper limit for low carbon steels should be 2.5×10^4.

A typical section of a friction weld, between solid bars, with upset is shown in figure. The maximum extent of upset is termed as the Apparent Upset while the biggest welded diameter decides the extent of Real Upset.

With reference to the above figure these values can be expressed by the following relationships:

$$\text{Apparent upset, } U_A = \frac{D' - d}{d} \times 100\%$$

$$\text{Real upset, } U_R = \frac{D - d}{d} \times 100\%.$$

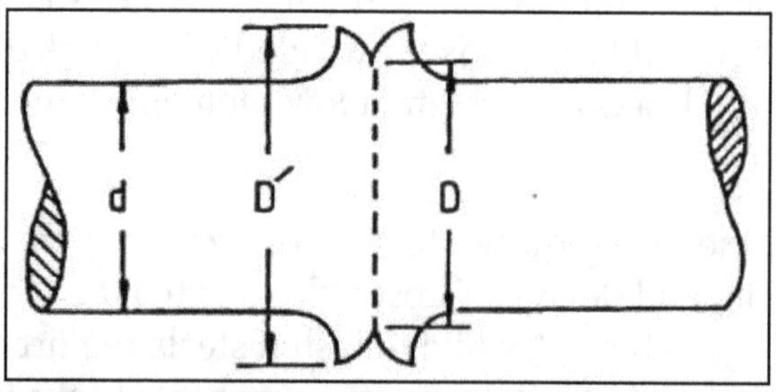

Figure: Configuration of a friction welded bar.

Fast braking is provided to stop the rotation quickly at the end of specified heating time or after a designed amount of axial shortening of the weldment. This provides desired control of overall weldment length and broadens the acceptable range of welding variables for critical applications. With small diameters, where the speed is high, very fast braking is required and this can be achieved by electric clutches, motor braking or friction braking. Also, arrangement can be made to release the originally stationary specimen towards the end of the heating cycle.

There are essentially two types of machines:

- Low power machines with power ratings of the order of 12 W/ mm², and
- High power machines with 35 to 115 W/mm² of the material welded.

If adequate pressure is available, it is possible to increase the capacity of a machine by using a bevelling technique, as shown in figure.

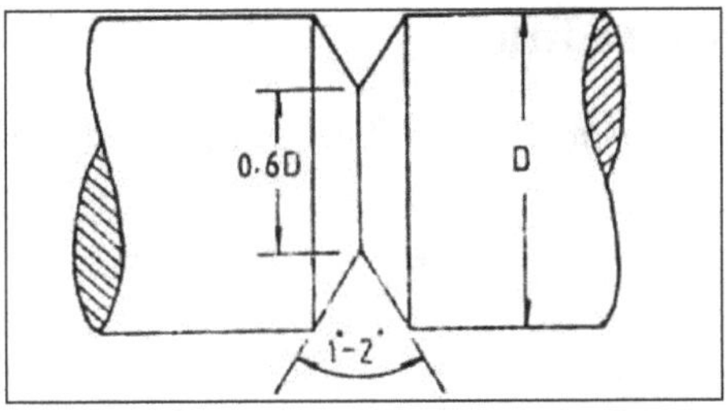

Figure: Use of bevelled bars to increase machine capacity.

Problem:

Find the power rating of a machine to friction weld a high tensile steel (0.2%C, 1%Cr, 0.4%Ni) having a forging temperature of 900 °C and a strength at that temperature of 125 N/mm^2. With a rotational speed of 3000 rpm, an upset gather of 2.8 mm in 10 mm diameter material is allowed. Assume that the coefficient of friction, μ= 1, and that under continuous shear, the shear strength of the material is equal to the compressive strength and that the torque acts at ⅔rd the radius of the workpiece.

Solution:

$$L = 125 N/mm^2; \qquad N = 3000 rpm$$

$$\mu = 1; \qquad d = 10 mm$$

Friction force/area, F = μL

$$= 1 \times 125 = 125 N/mm^2$$

$$\text{Torque, } T = \text{force} \times \text{area} \times \text{torque arm}$$

$$= F \times A \times \left(\frac{2}{3}r\right)$$

$$= F \times \left(\frac{\pi}{4}d^2\right) \times \left(\frac{2}{3}r\right)$$

$$= 125 \times \frac{\pi}{4} \times 10^2 \times \frac{2}{3} \times 5$$

$$= 32738 \text{ N-mm}$$

$$= 32.738 \text{ N-m}$$

$$\text{Power of machine, } p = \frac{\text{torque} \times \left(\frac{2\pi \cdot n.}{60}\right)}{1000} \text{KW}$$

$$= \frac{32.738 \times \left(\frac{2\pi \times 3000}{60}\right)}{1000} \text{KW}$$

$$= 10.3 \text{ KW}$$

Variables of Friction Welding

Three major variables in continuous drive friction welding are:

- Rotational speed.
- Axial pressure.
- Heating time.

1. Rotational Speed: Rotational speed provides the necessary relative velocity at the faying surfaces. Its magnitude depends upon the metal being welded and for steels, the tangential velocity for both solid and tublar workpieces, should be in the range of 75 – 110 m/min. Tangential s ds lower than 75 min result in excessive torque with consequential clamping problems, non-uniform up-set and tearing of metal at the joint. Friction welding machines for production purposes, handling 50 to 100 mm diameter workpieces, usually operate at speeds varying between 90 to 200 m/min.

High rotational speeds are useful for welding but axial pressure and heating time must be carefully controlled to avoid overheating of the weld zone particularly for welding quench hard-enable steels to control cooling rate and possible cracking.

In dissimilar metal welds, low rotational speeds can minimise the formation of brittle intermetallic compounds; however in general for controlling weld quality, rotational speed is not considered as a critical parameter.

2. Axial Pressure: The applied axial pressure controls the temperature gradient in the weld zone, the power required for the machine, and the axial shortening of the workpiece. The specific pressure depends upon the metal being welded and the joint configuration. It can be used to compensate for heat loss to a large body as in the case of tube-to-tubeplate welds.

Applied pressure must be high enough during the heating phase so as to keep the faying surfaces in close contact to avoid oxidation. Joint properties may often be improved if the applied pressure is increased at the end of the heating phase.

For making sound welds in mild steels the heating pressure used is usually 30 to 60 N/mm^2 while the forging pressure may lie in the range of 75 to 150 N/mm^2, and the commonly used values are 55 to 135 N/mm^2. However, higher forging pressures are required for high hot strength alloys like, stainless steels and nickel base alloys. If a preheating effect is required than the axial pressure of 20 N/mm^2 is applied initially for a short period which is then raised to the normal heating pressure.

3. Heating Time: Heating time is controlled depending upon whether a fixed pre-set time is allowed for healing or the extent of the axial upset is to be within the specified limits.

Excessive time limits productivity and results in wastage of material; while insufficient time may result in uneven heating as well as entrapped oxide and unbonded areas at the interface. The duration of welding for a 25 mm diameter bar, should be from 5 to 7 seconds at a rotational speed of 100 0 rpm.

Weld Properties of Friction Welding

One of the attractive features of friction welding is the metallurgical quality of the welds; the rapidity of heat generation produces an almost negligible heat affected zone. Due to good control of temperature and as the plastic metal is subjected to hot working during the heating phase, and to cold working during the forging phase, these results in a weld with extremely fine grain structure.

Metallurgical examination shows no evidence of melting as the measured temperatures for steels are usually in the range of 1260 to 1330 °C. However, rapidity of friction welding leads to high cooling rates which results in higher hardness of weld zone. Friction welds in hard enable steels, therefore, most often have to be annealed after welding.

Many times friction welding is employed to join dissimilar metals with a view to economising in the use of expensive alloy steels and high temperature alloys. It has been shown that 18/8 (Cr/Ni) stainless steel is actually work- hardened from 200 to 250 VHN on the stainless steel side due to the diffusion of carbon in it.

In a weld between 18/8 stainless steel and 20% Cr-Mo steel, the hardness of Cr-Mo steel is increased from 175 to 405 VHN but it can be reduced to 250 VHN by annealing. The increased hardness of stainless steel, however, remains unaffected by annealing.

Satisfactory welds between aluminium and stainless steel can be produced without the formation of brittle intermetallic compound layer. However, welds between aluminium and mild steel, and aluminium and copper may result in the formation of intermetallic compounds at the interface which may be reduced in the latter case by increasing the forging pressure to about 200 N/ mm^2.

Joint Design for Friction Welding

Basic joint design for friction welding is the same as for flash butt welding i.e., as far as possible like areas should be welded as shown in figure. The difficulty of welding two unequal sections arises due to different heat sinks on either side of the joint resulting in uneven heating and upsetting. In the case of a joint between a rod and a plate of the same material the plate thickness should be one-quarter of the rod diameter.

It is not possible to butt-weld two square bars of the same section to each other as it results in exposure of hot metal and consequent oxidation; however a large hexagonal bar may be welded to a smaller circular bar as there is no exposure of hot metal in such a case.

For successful friction welding, the outer diameter of the workpiece should not exceed the other by more than 1.33 times. The length projecting from the chuck should be 20-25 mm. The clamped length of the component should not be less than the weld diameter. When bill's or tubes are welded to plates most of the material that forms the flash comes from the bar or lube; this is because there is less mass in the smaller section and therefore heat penetrates deep into it.

For conical joints the faces are bevelled so as to have an angle of 45° to 60° to the axis of rotation, larger angles are preferred for low strength metals so as to support the axial thrust required to produce adequate heating pressure.

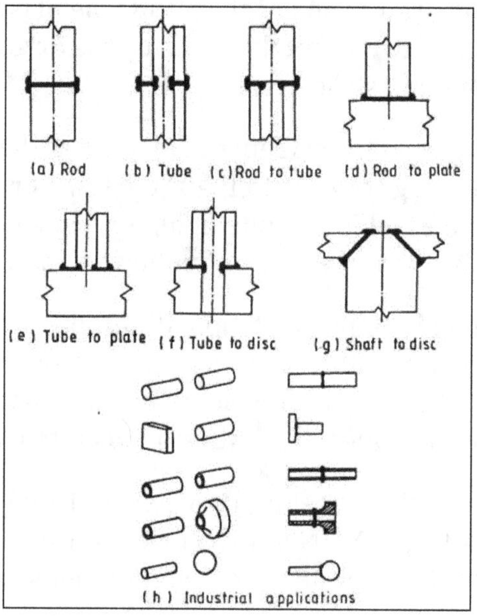

Figure: Typical joint designs and some industrial applications friction welding.

Welding of dissimilar metals can be eased by ensuring that both parts deform similarly. Similar degree of deformation can be facilitated by preheating the harder component by friction heating against an auxiliary plate which is removed at the appropriate moment Gas burners or high frequency induction heating can also be employed for the purpose. Still another method is to use a collar or a holder with an internal bevel which is placed around the soft stationary component in order to contain and direct it towards the harder material as shown in figure.

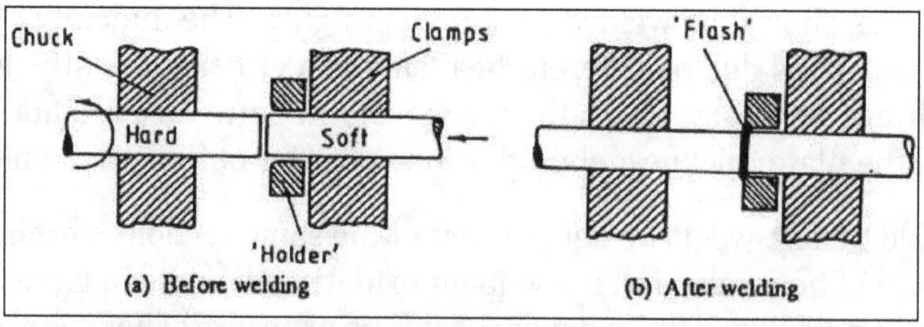

Figure: Technique for controlled shape of upset metal in friction welding.

Friction welding of dissimilar metals with widely different mechanical or thermal properties can be facilitated by having larger surface area for metals of lower strength or lower thermal conductivity. When flash cannot be removed conveniently clearance for it can be provided in one or both components.

Applications of Friction Welding

Friction welding is frequently being used instead of flash or upset welding for applications where one of the components to be joined has axial symmetry. Compared with flash welding friction welding has the advantage of cleanliness and balanced steady load on the mains; also it can be installed along with other machine tools and can be easily automated for high rate production.

Almost any metal that can be hot forged and is unsuitable for dry bearing application can be friction welded; however some metals may require post-weld heat treatment to remove the effect of quench hardening at the weld interface. Free machining alloys are difficult to friction weld as they often result in welds with planes of weakness in weld zone due to the re-distribution of inclusions. Such welds often exhibit lower values of strength, ductility and notch toughness.

Friction welding can be employed to weld solid material from 5 mm to 100 mm diameter or the equivalent areas in tubes and pipes. Due to solid- state joining friction welds have high quality performance both in similar and dissimilar combinations.

Carbon steels upto 1.1% C can be easily welded with the exception of free-cutting variety. Ni-Cr steels upto 18% Ni and 8% Cr and dissimilar steels can be easily welded; the range may include welding of 18/8 stainless steel to $2^1/_4$% Cr-Mo steel.

Welding of steel, due to its lower conductivity and greater plastic range, is comparatively much easier than the welding of non-ferrous metals and their combinations. The largest single user of friction welding is the automobile industry for volume production of components like manufacture of axle casings for cars and heavy vehicles; the joint produced is shown in figure. Another important use is the manufacture of twist drills in which tungsten high speed steel ends are welded to carbon steel shanks.

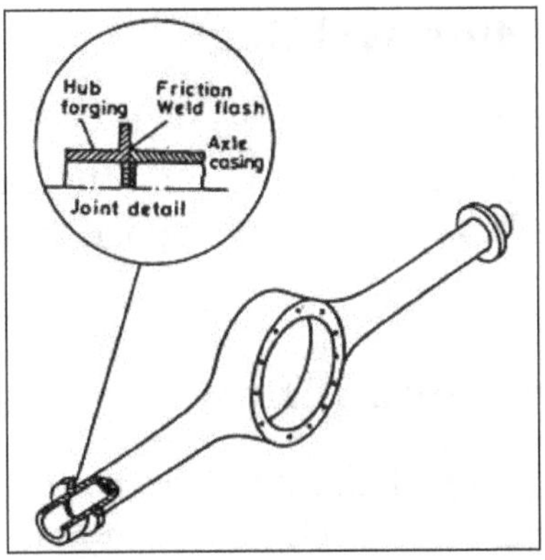

Figure: A typical friction welded hub-end forging to the axle casing for a heavily vehicle.

One of the important applications of friction welding is the welding of studs to plates of any thickness; another application of the process is the production of marine engine valves, the valves so produced are as good as or superior to those produced by forging. Impeller for the turbo-supercharger of a diesel engine may be made by friction welding of a carbon steel shaft to investment cast heat-resisting austenitic steel.

End sealing of tubes, as shown in figure, can also be achieved by friction welding. Dissimilar welded combinations may include joining of stainless steel to zirconium. Figure shows the edge preparation for joining stainless steel tube to a zirconium bar. For successful joining of dissimilar metal combinations it is essential to use high rotational speeds (more than 3200 rpm) so as to reduce the thickness of the intermetallic zone to the minimum.

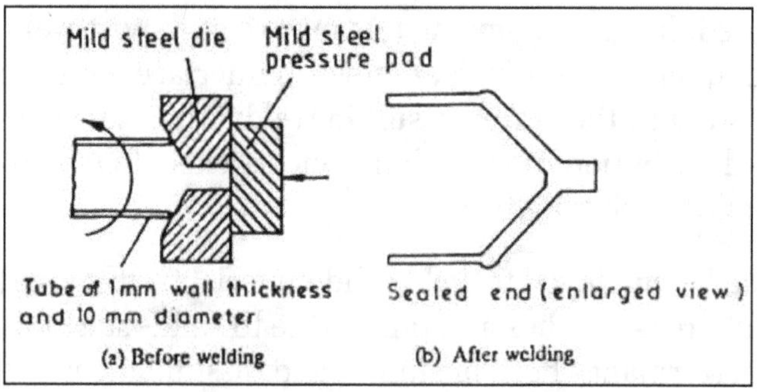

Figure: A setup for friction-welded end sealing of tubes.

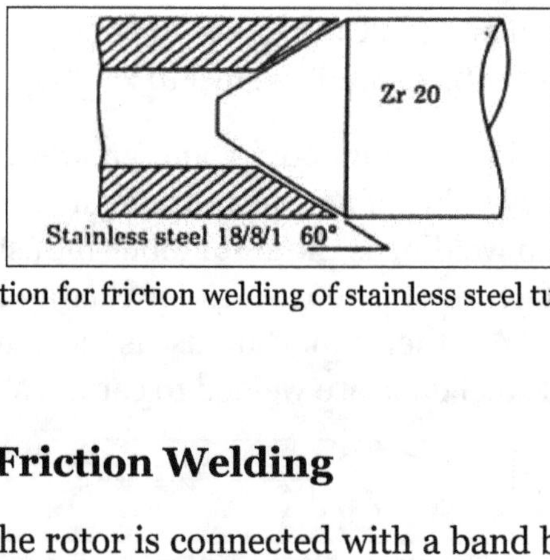

Figure: Edge preparation for friction welding of stainless steel tube to ziroconium bar.

Types

Continuous Induce Friction Welding

In this welding process, the rotor is connected with a band brake. When the temperature reaches the limit of plastic temperature, then the band brake comes into action whose role is to stop the rotor but the pressure applied on the workpiece increases until the weld is formed.

Inertia Friction Welding/Spin Welding

In this welding process, in place of the band brake, there are engine flywheel and shaft flywheel. These flywheels connect chuck to the motor. Firstly both flywheels are connected with each other. When the speed or friction reaches its limit, the engine flywheel gets detached from the shaft flywheel. Shaft flywheel has a low moment of inertia which stops automatically without any help of brake. The pressure force is continuously applied to the workpiece until the weld is formed.

Linear Friction Welding

This process is same as the spin welding process but in this process, the chuck rotates circularly whereas in linear friction welding it oscillates instead of spinning. It has a much slower speed as compared to spin welding which is beneficiary for job in linear friction process because many times the quality of the weld is better than that of spin welding.

Linear Vibration Welding

In linear vibration welding, the materials are placed in contact with each other and put under high pressure. An external vibration force is then applied to rub the pieces against each other, perpendicular to the pressure being applied. Both the workpiece are vibrated at 200 Hz frequency. Mostly this technique is used in the automotive industry.

Friction Surfacing

This process is derived from a friction welding process where a coating material is applied to a substrate. Basically it is a surface coating process. A rod composed of the coating material which is called mechatrode. Then it is rotated under pressure, generating a plastic-type layer in the rod at the interface with the substrate. By moving a substrate across the face of the rotating rod a plastic-like layer is deposited.

Advantages

- It can produce high quality of welds in a short period of time.
- No filler metal and flux is used in this process.
- It consumes less time than any other process.
- In this process, less skill is required to use it.
- Less area is affected by the heat produced in the process.
- A wide variety of metals can be operated on this system.

Disadvantages

- This process is limited to angular and flat butt joints.
- The preparation of the workpiece is more critical than making a weld.
- The cost of the setup of the machine is very high.
- It can only be used for smaller parts of machines; big parts are not compatible with it.

Explosive Welding

Explosive bonding is considered to be a solid-state welding process that uses controlled explosive energy to force two or more metals together at high pressures.

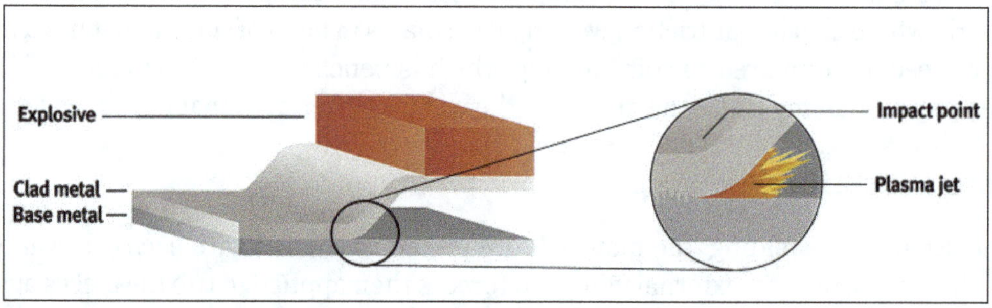

The resultant composite system is joined with a high-quality metallurgical bond. The time duration involved in the explosive welding event is so short, that the reaction zone between the constituent metals is microscopic. During the bonding process, several atomic layers on the surface of each metal become plasma. The collision angle between the two surfaces (typically less than 30°) forces the plasma to jet ahead of the collision front, effectively scrubbing both surfaces and leaving virgin metal.

The remaining thickness remains near ambient temperature and acts as a huge heat sink. Therefore, the bond line is an abrupt transition from the clad metal to the base metal with virtually no degradation of their initial physical or mechanical properties. The obvious benefit from this process is the joining of metallurgically incompatible systems. Any conventional cladding method, which uses heat, may cause brittle inter-metallic compounds to form.

Metallic Bond

The welded joint produced with the help of explosives is based on a metallic bond. The question arises, however, of what actually constitutes a metallic bond. Each metal exists as a result of this bond. A metallic bond is produced thanks to valence electrons that bind metal atoms together. These valence electrons are able to move freely through the entire metal lattice forming a shell or cloud together. This is sometimes also known as a gas cloud of free electrons, which is also the reason why a metal is able to conduct an electric current. Naturally, the atoms retain their fixed electrons around the atomic nuclei, which, by definition, are called non-valence electrons.

During the explosive welding process, a powerful shock drives the metal atoms into each other's sphere of influence, forcing the valence electrons into each other's paths. This is essentially how a metallic bond is created, the properties of which, in theory, are not inferior to those of a normal metallic bond.

Explosive Welding Process

In order to obtain a good idea of the processes involved, we will take the example of a carbon steel sheet which needs to be clad with a high-quality metal. After the characteristics of the materials to be clad have been assessed, the sheets are stripped of their oxide layer by means of an automated scouring process. They are then placed on top of one another at a specific distance using plastic spacers. A frame is then fitted on top of the upper sheet which will contain the powder explosives. A detonator is attached and the whole pack is ready for the explosive welding process. Activation of the detonator triggers a powerful explosion that only lasts a fraction of a second. Although different theories exist with regard to what actually happens during this explosion, the opinion below seems to be one of the most logical and plausible.

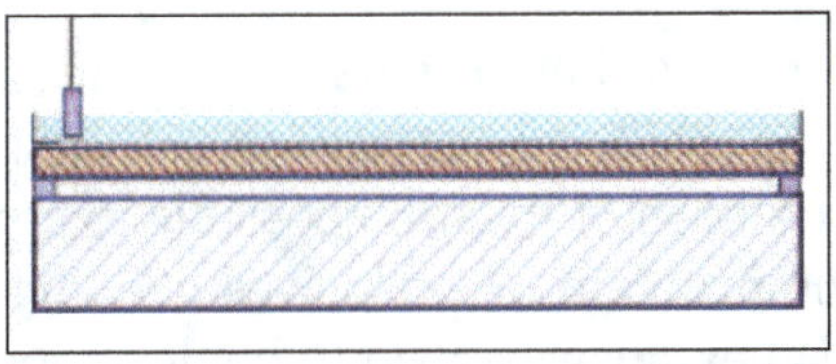

Figure: Illustration of the set-up needed for the explosive welding process.

The highly explosive force drives the top plate onto the base plate at an extremely high velocity creating a collision point. At the point of collision, the pressure is so high (several thousand bars) that the metals become super-plastic in highly localised areas. There are even indications of some atomic layers briefly turning to fluid due to this enormous pressure, even though the increase in temperature is negligible.

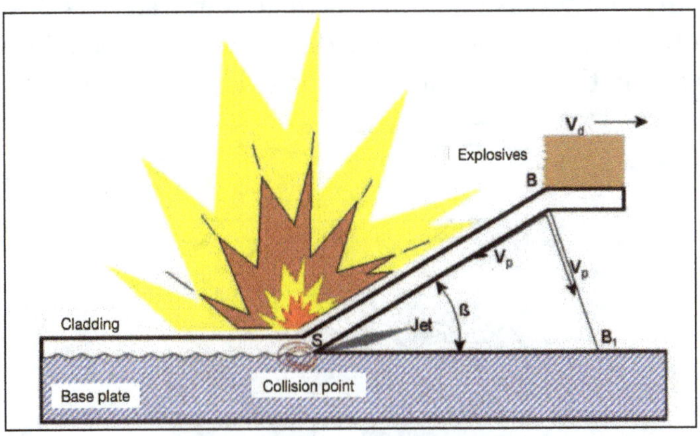

Figure: Explosive welding process.

During the explosion, a shockwave is created which is transmitted through the two metal layers in a similar fashion to ripples in water. Due to the superplasticity of the metals in the area of collision, the shockwave is able to produce a slight wave effect in the material. This wave pattern can often be seen with the naked eye where the metals join. The high velocity of this wave effect strips away any newly formed oxides on both surfaces. This can best be compared to a rug that is shaken out forcefully. The detached oxides are expelled as a flow of matter just ahead of the detonation front. This flow of matter is also known as a jet. This action leaves behind clean metal surfaces which have no time to oxidise, thereby allowing a metallic bond to be created. The bond is produced as a result of the high pressure which is still present and which causes the valence electrons to cross into each other's sphere of influence. This is how the metallic welded joint is eventually achieved. This welding process is also known as a variation of the cold pressure welding process as the entire process is carried out at ambient temperature. It will also be evident that the air between the sheets is expelled during the cladding process unless this process is conducted in a vacuum environment.

Open Air Cladding versus Vacuum Cladding

Explosive welding or cladding can therefore be carried out in the open air or in a vacuum environment. The latter is easier said than done as it first requires a special vacuum dome to be developed. In addition, it must be possible to create a sufficient vacuum in this space. There is also the problem of explosives needing oxygen during the explosion. This means that special explosives need to be developed that generate their own oxygen when activated. It is clear that vacuum cladding is a far more complex process than atmospheric cladding. The question therefore arises as to why carry out this process in a vacuum environment at all. The answer to this is that it mainly has to do with the superior quality of the welded joint achieved in the end. There are also additional advantages such as being able to work regardless of the weather and a significant reduction in noise as noise requires air in order to travel. Yet another advantage is that fewer explosives are required as no air needs to be removed. Besides saving on explosives, there is another benefit which is related to the type of transition. In practice, the shockwave can result in three types of transitions - laminar, wavy or turbulent.

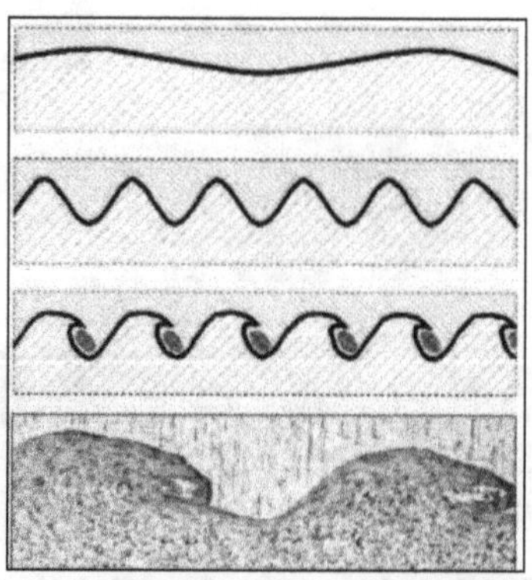

Figure: Three types of transitions resulting from explosion welding - laminar, wavy and turbulent. Sometimes the waves can even tip over.

If an inexperienced person were to look at these transitions, they would quickly come to the conclusion that the turbulent bond could well be the most optimum as the metals are noticeably interlocked. This would be a mistake, as the bond is not created by an interlocking process or something similar after all, In fact, the most attractive transition is a wavy one, but in practice the immense speed of the process often causes the peaks of the waves to top over creating a turbulent transition. This can somewhat be compared to waves in the sea which curl over when they reach a certain height. In other words, the greater the shockwaves, the greater the chance of a turbulent transition being created.

When the super-plastic waves curl over, some of the oxides from the oxide flow (jet) that is created become trapped. This leads to the development of unwanted oxide pockets or conglomerates in the wave patterns. These pockets are visible to the naked eye. If a clad strip of this type were to be bent, these oxide conglomerates would open out even more. In theory, acceptable mechanical values can be achieved with these types of transitions and their shear values can be even better than for a laminar transition.

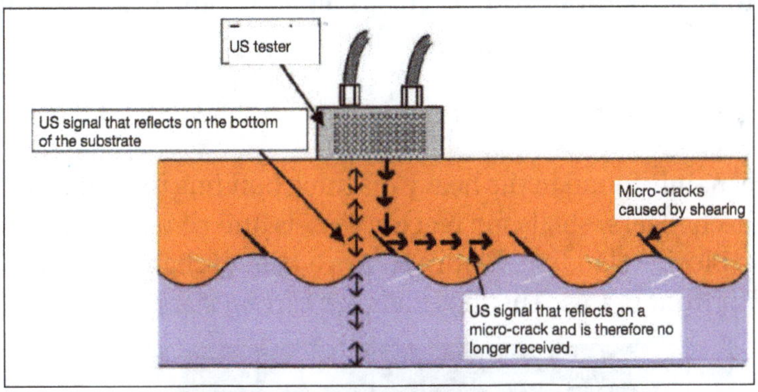

Figure: US signals are dispersed by the micro-cracks that can develop on the peaks of the waves.

The disadvantage however, is that foreign particles have become trapped, which can undermine resistance to corrosion in certain applications. The oxide pockets are porous and are also hygroscopic which, partly due to the potential difference of the two clad metals, can accelerate the development of corrosion. This is especially the case with clad connecting strips which are used to weld an aluminium deckhouse to the steel deck of a ship. Another disadvantage is that high wave patterns can produce micro-cracks on the peaks of the waves. These are caused by internal shearing in the metal lattice. The risk is particularly high with less ductile metals.

Internal research has therefore shown that the laminar transition provides the best connection as it offers the following advantages:

- These types of transitions remain free of oxides and any brittle intermetallic connections that can develop due to extreme pressures as is the case when using a lot of explosives.
- It has also been shown that microcracks can develop on the peaks of waves in a high wave pattern; particularly when metals are less ductile. This is illustrated in figure. An ultrasonic tester can be used to determine these cracks. This is a problem that does not arise with laminar transitions.

- As fewer explosives are used for laminar transitions, there is also considerably less strain hardening of the metal surface. It can actually be said that the structure of the two metals remain unchanged as do the other initial properties. It is evident that this also offers a significant advantage.

Fewer explosives are therefore required for vacuum cladding, which means that, in principle, a laminar or slightly wavy transition is achieved. No oxide pockets develop in this case as these oxides are able to escape in time. An additional advantage is that tests have shown that fatigue strength is also superior. This is why the offshore industry, in particular, prefers using vacuum clad connecting strips to strips produced in the open air. This also applies when these strips are used asymmetrically in ship building as they are better able to absorb torque forces. This is all due to the absence of micro-cracks in the wave peaks.

Applications

Applications are mainly found in the heavy machine building industry. In particular, tube sheets for heat exchangers, shells for reactor vessels and so on can be manufactured effectively using explosive cladding. Figure shows a heat exchanger with a clad tube sheet and a clad shell in front of this. The materials are carbon steel clad with super duplex.

Figure: A heat exchanger with clad tube sheet and shell. The base is carbon steel and the cladding is super duplex.

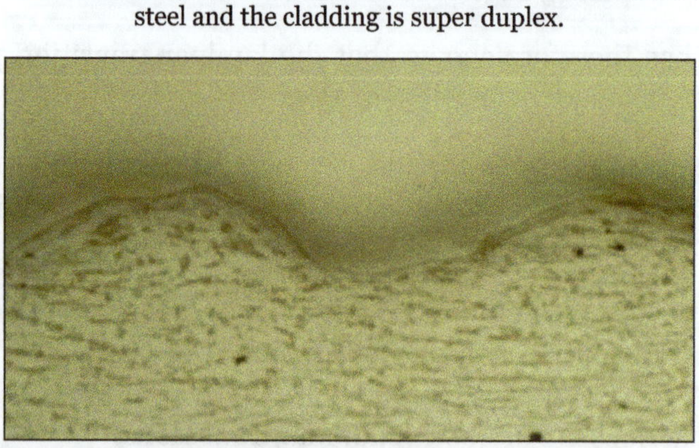

Figure: Perfect waving interface without any turbulence or oxides.

Further, this type of joint can also serve as a starting point in the manufacture of roller clad plates. It is also used to make transition joints which are sawn from sheets as strips. These strips can be used in the shipbuilding industry, for example, to weld an aluminium superstructure to steel decks. Not only does this save on rivets or nuts and bolts but also minimises the risk of corrosion.

Electrolysis companies that reduce primary metals from ores often use aluminium anode rods that need to be connected to steel or copper. Aluminium is a good conductor of electricity but as the temperature in the vicinity of the electrolytic cell is too high, a steel frame connected to the carbon block is required in this area. By welding a cladding between the steel and aluminium, there will be no transition resistance.

If aluminium is connected to aluminium mechanically, then this results in a transition resistance because of the oxide layer. Copper does not have this and this can therefore lead to substantial savings on energy bills. These types of connections are preferable as these contacts must be detachable.

Solid-state Diffusion Bonding

In diffusion bonding or diffusion welding a weld is achieved by the application of pressure, of the order of 5 to 75 N/mm^2, while the pieces are held at a high temperature, normally about 70% of the melting point in degrees absolute i.e. about 1000 °C for steel. The process is based on solid-phase diffusion which, obviously, is accelerated with rise in temperature.

Diffusion in metals takes place due to vacant lattice sites or along grain boundaries, and is expressed by the following mathematical relationship:

$$D = D_0 e^{-ERT}$$

where,

D = rate of diffusion.

D_0 = Constant having the same dimension as D,

E = activation energy,

R = gas constant,

T = Absolute temperature at which the workpieces are held.

Depending upon the extent of diffusion required the process may be completed in 2 to 3 minutes or may take many minutes or even hours. The quality of surfaces to be welded plays an important role. A good quality surface turned, milled or ground to a standard of 0-4 to 0-2 μm* CLA (centreline average) is usually adequate. The surface must be degreased before welding by using acetone or petroleum ether swab.

Presence of oxide layers on the surfaces being joined hinder diffusion but get dissipated over a period of time. Thus, metals which dissolve their own oxides such as iron and titanium, bond easily. On the contrary metals that form tough refractory oxide layers, like aluminium, are difficult to diffusion weld.

Diffusion bonding can be achieved by three methods viz:

- Gas pressure bonding.
- Vacuum fusion bonding.
- Eutectic fusion bonding.

In gas pressure bonding, the parts are held together in an inert atmosphere and heated to a temperature of 800 °C by a system resembling an autoclave. During heating the high pressure provides uniform pressure over all the surfaces. This method is used for bonding non-ferrous metals only because it necessitates high temperatures for steels.

In vacuum fusion bonding the parts are held in an intimate contact in a vacuum chamber. The pressure on the parts is applied by mechanical means or a hydraulic pump, and heating is done in the same way as in gas pressure welding. Figure shows a schematic diagram for vacuum fusion bonding.

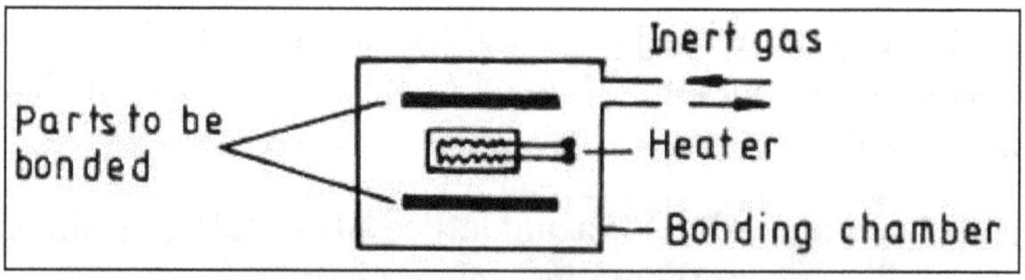

Figure: A setup for diffusion bonding.

A vacuum pumping system which can quickly reduce pressure to about 10^{-3} torr (mm of mercury) needs to be used. High pressure created by the use of mechanical or hydraulic means makes it possible to diffusion bond steels by this method. Successful joining of steel can be achieved at a temperature of about 1150 °C under an applied pressure of nearly 70 N/mm^2.

In eutectic fusion bonding a thin piece of a particular material is placed between the surfaces to be welded. These results in the formation of a eutectic compound by diffusion at an elevated temperature and the piece may completely disappear and form eutectic alloy at the interface. The material used for being placed in between the two parts is usually a dissimilar metal in foil form with a thickness of 0-005 to 0-025 mm.

Diffusion bonding can be used to join dissimilar metals e.g., steel can be welded to aluminium, tungsten, titanium, molybdenum, cermet's (compounds of ceramics and metals), copper to titanium, titanium to platinum, etc. It finds use in radio engineering, electronics, instrument making, missile, aircraft, nuclear, and aerospace industries.

Typical applications of diffusion bonding include tipping of heavy cutting tools with carbide tips or hard alloys, joining of vacuum tube components, fabrication of high temperature heaters from molybdenum disilicide for resistor furnace that can operate in an oxidising atmosphere upto 1650 °C. In aerospace industry it is used for fabricating complex shaped components of titanium from simple structural shapes. It is also used for surfacing components to resist wear, heat or corrosion.

Principle and Working

This process works on basic principle of diffusion. Diffusion means movement of molecules or atoms from high concentration region to low concentration region. This is fundamental principle of diffusion welding. In this welding process both the welding plates are placed one over other in high pressure and temperature for a long period of time. This high pressure force starts diffusion between interface surfaces. This diffusion can be accelerated by the application of high temperature. This temperature does not melt the welding plates. The temperature range is about 50-60% of melting temperature. This whole process takes place in vacuum or in inert environment which protects the welding plates form oxidation.

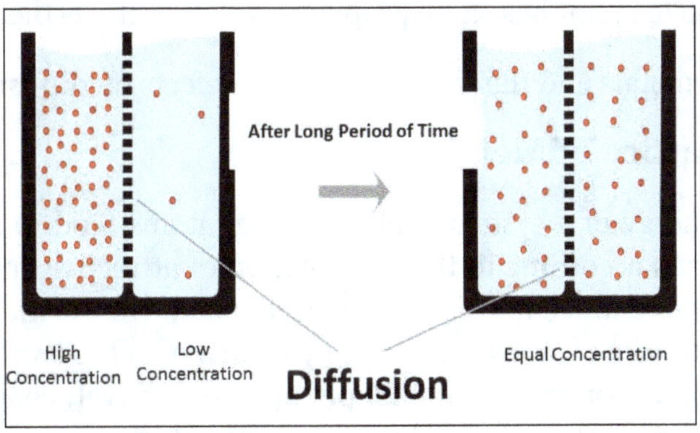

The working of diffusion bonding can be summarized as follow:

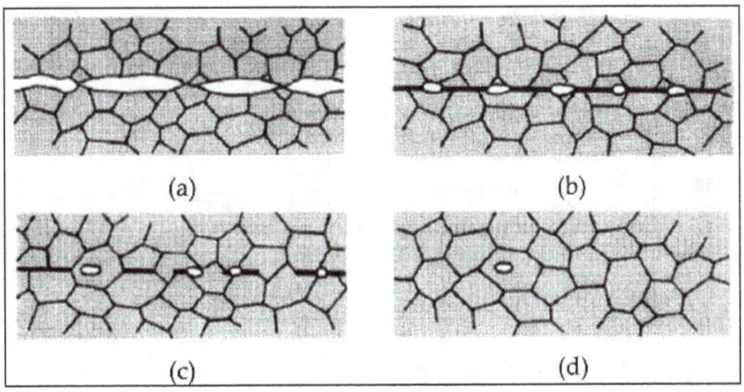

- First both the welding plate surfaces prepared for welding. In this process, both the interface surfaces made equally flat which is basic requirement of diffusion process. The interface surfaces should be machined, cleaned and polished well which remove all chemical contaminants from the surface. Any contaminant particle can be reduced diffusion between welding plates.
- Now both the plates are clamped and placed one over another. This assembly placed into a vacuum chamber or in an inert environment. This protects the welding surface from oxidation.
- A high pressure and temperature applied on this assembly to start diffusion. The temperature applied by the furnace heating or electric resistance heating. The high pressure is applied by a hydraulic press, dead weight or by the differential gas pressure. These conditions are maintained for a long duration of time for proper diffusion.
- At the starting stage of this process, local deformation at the interface surface due to creep and yield take place. Now the diffusion takes place which form an interface boundary.
- After a long period of time, both the plates properly diffused into one another which makes a strong joint. The interface boundary disappear which form a clean joint. This joint has same properties or strength as the base material.

Micro-structure of metals and the impact of lattice defects on diffusion welding:

Atoms in the Lattice of Metals

To minimise the energy of a system, isolated metal atoms tend to arrange in a regular lattice at positions according to the annihilation of attractive and repulsive forces. The positions are well-defined and specific of each metal. Hence, they can be used, e.g., for determining the composition of an alloy by means of WDX (wavelength dispersive X-ray). When forming a compound, atoms split up into positively charged atomic nuclei, while valence electrons are transferred to the so-called electron gas and can move freely within the lattice. Consequently, metals are good conductors of heat and electricity.

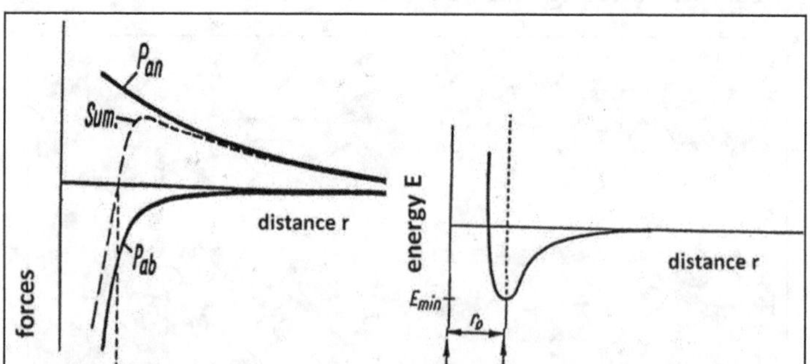

Figure: Equilibrium of attractive and repulsive forces in the metallic lattice.

Thermal Expansion

Depending on the thermal energy of the whole system, the positively charged atomic nuclei oscillate around their position, leading to a thermal expansion. According to Grüneisen's rule, linear expansion is in the range of 2% and volumetric expansion is 6–7% up to the melting point of a metal. Hence, the melting point can be used to estimate the thermal coefficient of expansion. Below the melting temperature, the oscillation amplitude is about 12% of the lattice constant.

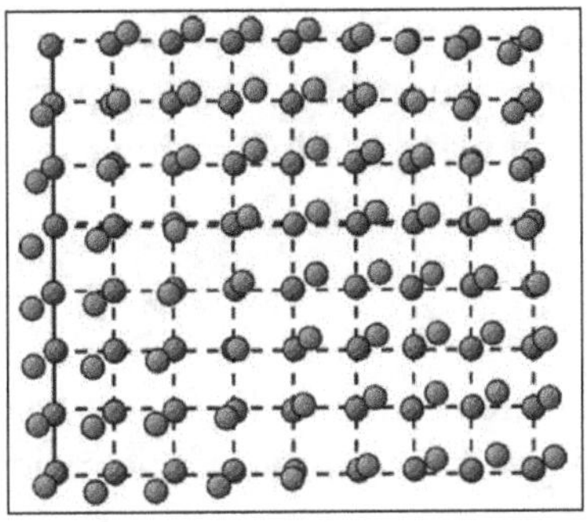

Figure: Thermal oscillation of atoms.

Thermal Activation, Diffusion, Polymorphism and Zero-dimensional Lattice Defects

With increasing thermal oscillation, not only the amplitude increases but also the energy of collisions between atoms. Gradually, some atoms are facilitated to leave its lattice sites and a vacancy is left leading to a punctual stress state. With increasing temperature, an exponentially increasing number of atoms is displaced from the lattice sites and the density of vacancies is considerably enhanced equation $c_V = \frac{n}{N} * \exp.\left(\frac{-\Delta U}{RT}\right)$

where, cV is the concentration of vacancies (cm^{-3}), n is the number of vacancies, N is the number of sites in the metallic lattice, U is the energy of formation of vacancies (for metals 80–200 J/mol), R is the gas constant (J/mol*K) and T is the temperature (K).

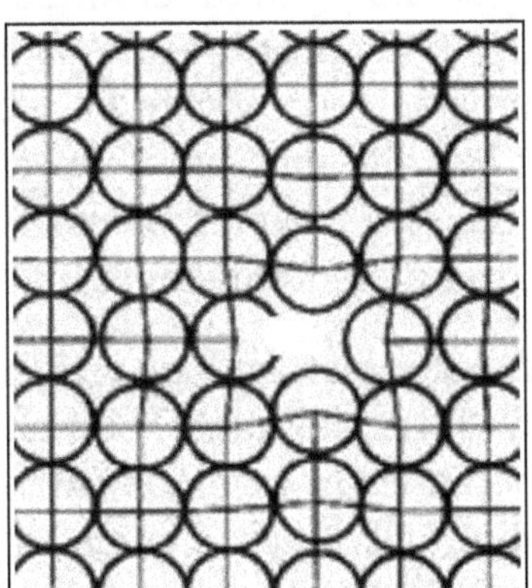

Figure: Vacancy in the lattice causing punctual stress.

Vacancies are regular lattice sites not occupied by an atom. Due to a missing atom, the surrounding atoms tend to fill the gap and the lattice is distorted at this point, representing a zero-dimensional defect.

The density of vacancies is 10^{-12} at room temperature and increases to 10^{-4} below the melting temperature.

Vacancies strongly facilitate the diffusion of atoms between different sites of the lattice and, hence, concentration facilitates the formation of a monolithic compound during diffusion welding. As a consequence, the coefficient of diffusion increases exponentially with temperature. An increase in bonding temperature by 20 K may result in a doubling of the diffusion coefficient, thus illustrating the strong non-linear influence of temperature on diffusion welding:

$$D = D_0 * \exp.\left(\frac{-\Delta U}{RT}\right)$$

where, D is the diffusion coefficient (m^2/s), D_0 is the frequency factor (material constant) (m^2/s) and U is the energy of formation of vacancies (J/mol).

The number of vacancies versus temperature can be plotted as a logarithmic function, the so-called Arrhenius plot.

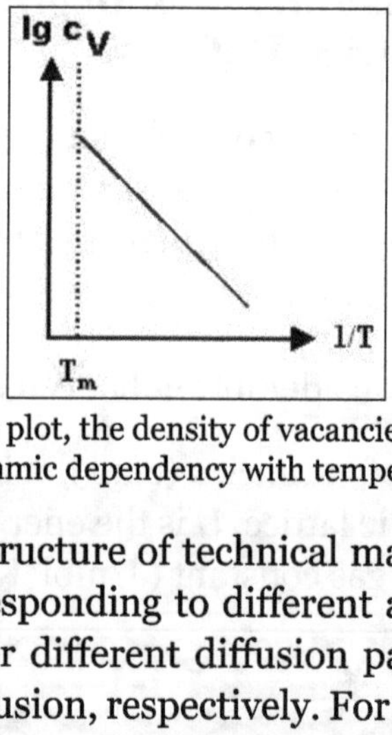

Figure: Arrhenius plot, the density of vacancies increases with a logarithmic dependency with temperature.

Depending on the real micro-structure of technical materials, different types of diffusion can be distinguished corresponding to different activation energies for different lattice defects. Straight lines for different diffusion paths can be plotted for surface, grain boundary and volume diffusion, respectively. For diffusion welding, grain boundary diffusion predominates at low and medium temperature. As the cross-section of grain boundaries is related to the volume and the density of vacancies increases exponentially, volume diffusion becomes predominant at high temperature.

At the same time, grain growth takes place at high temperatures, which minimises the interfacial energy of the system. If the material shows no polymorphic transformation or the grain boundaries are not pinned by insoluble intra-granular precipitations (e.g., for ODS alloys), diffusion welding will be accompanied by grain growth.

Technical materials are no pure metals, but also contain other sorts of atoms, e.g., alloying elements like manganese, chromium or carbon for steel. Similar to vacancies, these atoms are integrated into the basic lattice as zero-dimensional defects. If they form the same type of lattice (e.g., cubic face-or cubic space-centred), and if the difference in atomic radii is less than 15%, they can occupy regular sites of the host lattice. Small non-metallic atoms with an atomic radius smaller than 59% of the host atoms can be dissolved interstitially like carbon in iron.

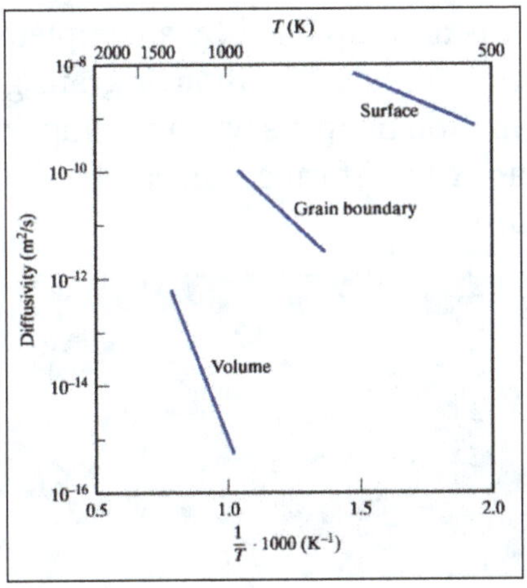

Figure: Different modes of diffusion of atoms versus temperature.

Although solubility of interstitial atoms is low, they have diffusion coefficients higher by some orders of magnitude in the lattice, since more suitable gaps are available. In case of low temperature rates during cooling down from diffusion welding temperature, this may be of relevance to the formation of undesired precipitations. The dwell time during diffusion welding should always be kept in the range of solution annealing for an alloy. This may conflict with a low temperature to limit grain growth.

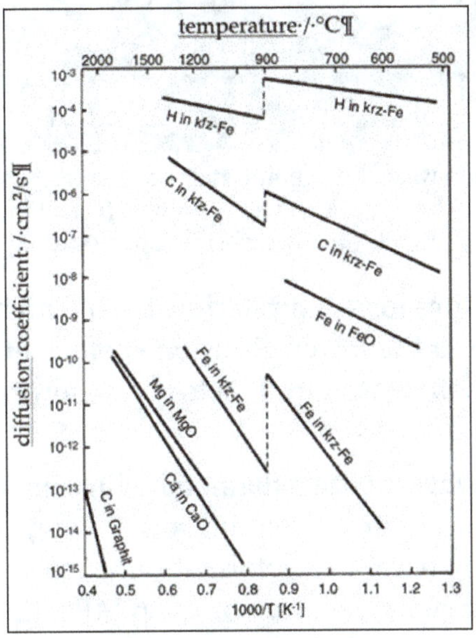

Figure: Diffusion coefficients of different sorts of atoms and depending on the type of lattice.

If a metal is polymorphic, abrupt changes in solubility and in the diffusion coefficient may occur. For iron, e.g., these parameters change by two orders of magnitude. Reasons are different solubilities for foreign atoms and different sizes of gaps between the atoms in the lattice. For example, maximum solubility of carbon in α-ferrite (cubic space-centred) is 0.02% at 723 °C, whereas the solubility of carbon in Y-ferrite (cubic face-centred) is 2.06% at 1143 °C, i.e., higher by a factor of about 100.

Additionally, polymorphism is accompanied by a complete new formation of the micro-structure, and grain size is reduced. Although grain growth occurs at high temperature, the same transformation happens when cooling down. This is the reason why normal steels, showing an a⇔Y transformation, can be diffusion welded easily with a finely grained micro-structure.

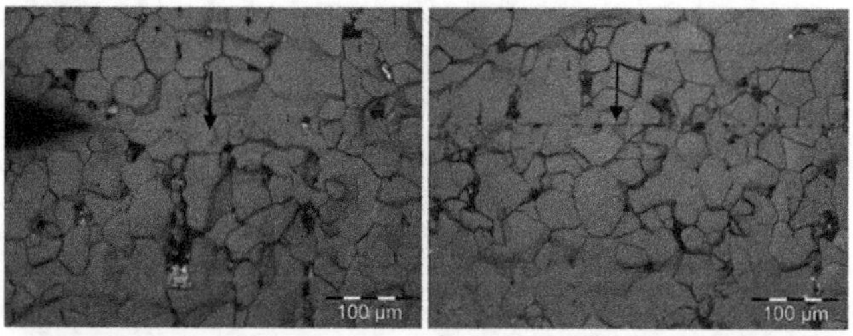

Figure: Diffusion weld of St 37 (1.0254), T = 1075 °C, t = 1 h, p = 10 MPa, normal α⇔γ-transformation, deformation: 3.13%.

On the other hand, a diffusion weld of austenitic steel is displayed in figure below. The impact of the four times longer dwell time on the grain size can be seen clearly.

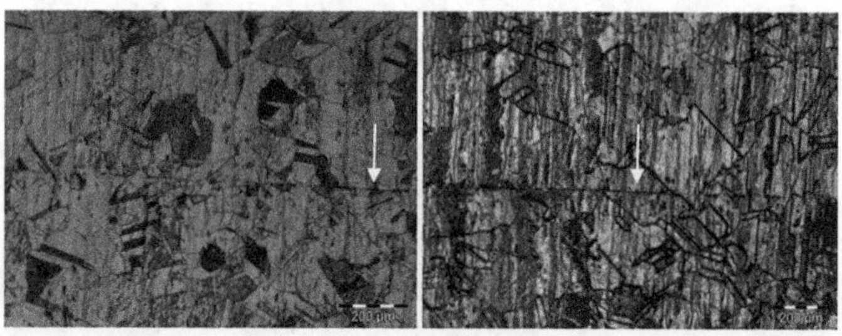

Figure: Diffusion weld of austenitic stainless steel AISI 304 (1.4301) at T = 1075 °C, p ≈ 15 MPa. Left: t = 1 h, deformation: 2.75%. Right: t = 4 h, deformation: 7.04%.

Although the bearing pressure for 1-h dwell time is 50% higher than for 1.0254, deformation is comparable due to the lower diffusion coefficient in the cubic face-centred lattice. When dwell time is increased by a factor of 4, however, deformation increases by a factor of about 2.5.

In Figure below, ten 1-mm layers with a diameter of 40 mm of a fully ferritic stabilised stainless steel were diffusion welded. Welding at T = 1075 °C, t = 1 h and p = 10 MPa for comparison failed due to excessive deformation. Even at a reduced temperature of T = 1000 °C and reduced bearing pressure of p = 6 MPa, the deformation was huge at 14.6%. For T = 950 °C, t = 1 h, p = 6 MPa, the deformation was still 3.8%.

These high deformations under these mild conditions have to be attributed to the high diffusion coefficient in ferrite. However, despite the high deformation and excessive grain growth due to lacking polymorphism, only very little grain growth across the bonding planes is visible in Figure, illustrating the role of surface passivation layers.

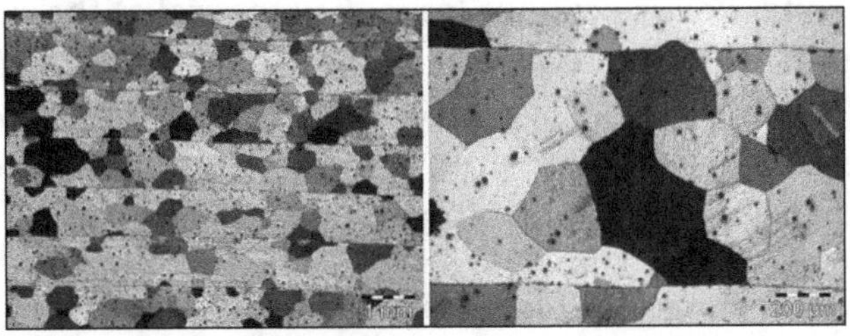

Figure: Diffusion weld of a fully ferritic stabilized stainless steel Crofer 22 APU (1.4760) at T = 1000 °C, t = 1 h, p = 6 MPa. Deformation: 14.6%.

One-dimensional Defects: Impact of Dislocation Density on Mechanical Properties

Dislocations represent an inserted plane in a metallic lattice.

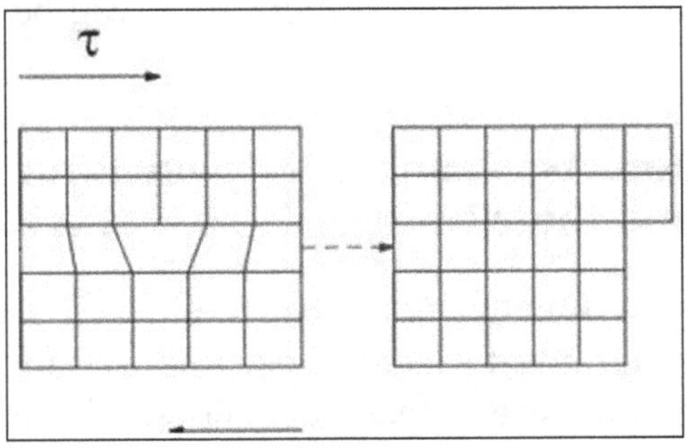

Figure: Plastic deformation by the movement of a dislocation across the lattice.

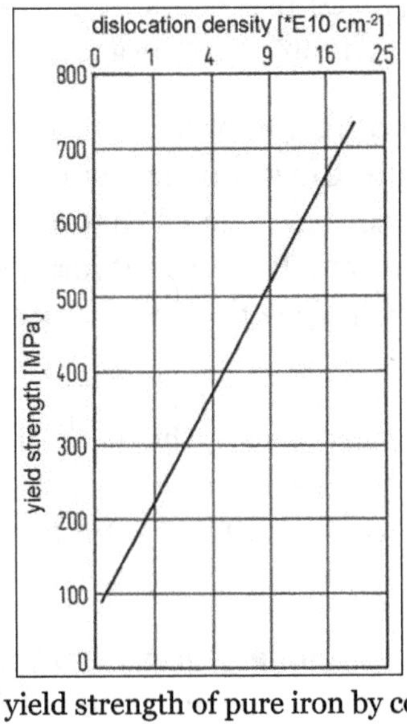

Figure: Increase of yield strength of pure iron by cold work hardening.

Of course, the inserted plane does not end at a constant level in different layers of the third dimension, but at an arbitrary depth, leading to complex stress conditions. If adequate shear stress appears, the dislocations are moved through the lattice in a step-wise manner, thus causing a plastic deformation. However, the dislocation density is not dropping, despite the dislocations leave the material at the surface. In opposite, it increases exponentially during cold working due to the so-called Frank-Read mechanism. Dislocation density can be given as a length of the dislocation line per unit of volume and can reach as much as 10^{12} cm^{-2}. Only by cold work hardening alone can the mechanical strength of a material be multiplied.

At room temperature, dislocation movement is the predominant deformation mechanism in metals. Dislocations represent a one-dimensional lattice defect.

Dislocations mean an energy excess compared to an undistorted lattice. Hence, at elevated temperatures of about 40% of the melting temperature of pure metals and 50% for alloys, recrystallization takes place. For metals showing no polymorphism, cold work hardening and subsequent recrystallization is the only way to reduce the original grain size. However, it is applicable to half-finished products only. Hence, when cold-worked material is diffusion welded, recrystallization will be included and affect the grain size.

Two-dimensional Defects: Grain and Phase Boundaries; Interfacial Layers and their Influence on Diffusion Welding

Two-dimensional defects of a metallic lattice are reflected, e.g., by grain boundaries. They can be described as an interfacial area per unit of volume and can vary over a wide range, whereas the grain size of technical alloys is in the range of about 5–200 μm. Coating technologies such as galvanic deposition, physical vapor deposition (PVD) or chemical vapor deposition (CVD) processes lead to amorphous or nanocrystalline micro-structures possessing a high internal energy.

Two-dimensional defects affect diffusion welding in several ways: first, the dislocation movement is limited according to the grain size, since grain boundaries are obstacles for movement through the lattice. This means that for a constant strain, deformation by dislocation movement will be smaller for a material with a small grain size. At elevated temperature, however, grain growth occurs and the driving force is larger for a fine-grained material.

A similar effect is observed when using the so-called nanofoils, a thin stack of multiple nanometre layers of different materials possessing a very high interfacial energy in a metastable state. As a result, very high temperatures can be achieved temporarily.

Secondly, deformation at elevated temperatures is governed by grain boundary sliding (Coble creep) or the flow of vacancies through the volume (Nabarro-Herring creep).

This means a coarsely grained material will tend to a larger deformation during diffusion welding because there are fewer obstacles for dislocation movement and grains tend to slide against each other.

It can be stated that the degree of deformation and the creep rate for a material during diffusion welding will depend on its grain size and will be very sensitive to the temperature used. More complex deformation behaviour may result from multiphase materials: phase boundaries can occur in a wide range of orders of magnitude, either between grains or within, e.g., as thin lamellas in grains of eutectic or eutectoid composition, such as perlite for steel.

Temporarily liquid phases (TLP) can be formed, e.g., by galvanic or PVD deposition of thin layers of two or more different metals, forming a low melting alloy during diffusion welding. Since the inter-layer diffuses into the bulk material, ideally a homogeneous material is left after finishing the process, which is insusceptible to inter-crystalline corrosion.

The opposite happens when different metals form inter-metallic compounds that are brittle and have a high melting temperature. In this case, bonding temperature and time should be limited, such that a thin layer only can be formed between both materials, which do not exhibit any excessive brittleness.

Three-dimensional Defects: Precipitation

As regards precipitations, it must be distinguished between soluble and insoluble species at diffusion welding temperature. Precipitation may be formed, e.g., due to a low cooling rate after diffusion welding in the range of solution annealing temperature. As a consequence, a two-phase micro-structure with coarse precipitations is formed at the grain boundaries. It is subjected to inter-crystalline corrosion. Examples are nickel-based alloys that lose their favourable corrosion resistance.

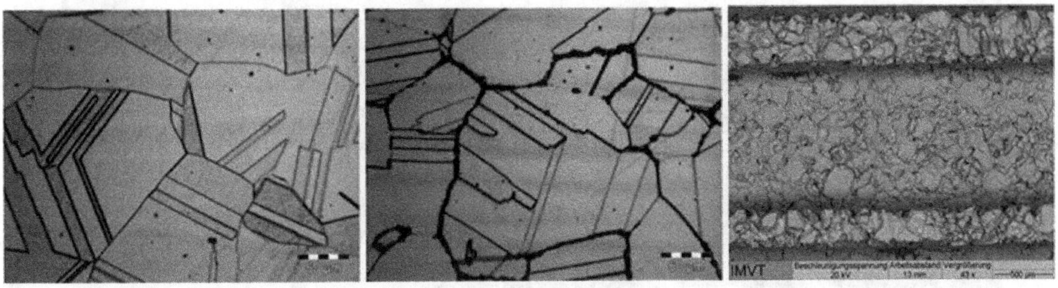

Figure: Micro-structure of Hastelloy C-22 (2.4602). Left: after quenching from 1100 °C/70 min in water. Middle: after cooling from 1100 °C with a rate of 3 K/min (1100 °C ≥ 650°C = 2.5 h). Right: corrosion attack after diffusion welding in 95–97% sulphuric acid at 100 °C and 1008 h.

Again, the size of precipitations determines the consequences. In case of nanoscaled, insoluble precipitations, e.g., for ODS materials, dislocation movement and grain growth is restricted very effectively.

For example, a pure OF-Cu showed good results at T = 850 °C. The dimension of the material was 28 × 15 mm², consisting of six micro-structured layers with a thickness of 3.04 mm and an overall height of 13.04 mm, respectively.

Figure: Diffusion welding of copper. T = 850 °C, t = 4 h, p = 2 MPa, micro-structured stack: 18.2%, overall: 4.2%.

Especially in thin-walled micro-structures, perfect grain growth across the bonding planes can be seen. However, in the massive border area, pores remain and grain growth is not as pronounced. The reason probably is a local excess of bearing pressure at the thin walls. However, in massive areas, the bearing pressure of 2 MPa is too low to deform asperities and fill pores sufficiently at this temperature.

Comparative diffusion welding experiments were made using two discs made of ODS copper Discup C3/80 with a diameter of 40 mm and an overall height of 6.88 mm. The surfaces were flycut using a polycrystalline diamond tool at a feed rate of 240 mm/min and 3000 rpm, giving a period of 80 µm feed per revolution at a very low roughness in the range of *Rt* = 1–1.5 µm, *Ra* = 0.2 µm. The roughness patterns of the discs were not aligned to each other.

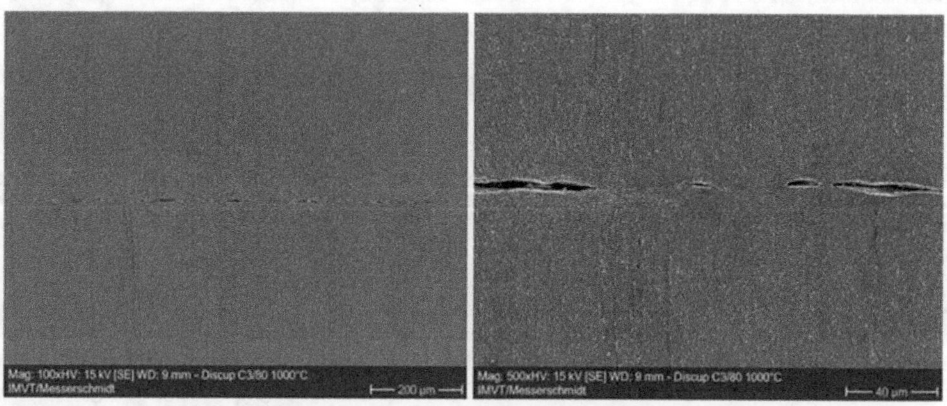

Figure: Diffusion welding experiment using Discup C_3/80, an oxide-dispersion-strengthened copper alloy. T = 1000 °C, t = 4 h, p = 6 MPa, deformation: 0.8%.

Discup alloys consist of pure copper containing a few tenths percent of sub-micron disperoids generated by reactive milling. Afterwards, the material is strongly deformed by extruding. The melting temperature is 1083 °C like for pure copper. Despite a much higher temperature and bearing pressure compared to the OF copper sample shown in Figure, the deformation is as low as 0.8%. SEM images are taken, illustrating very poor joining of the mating surfaces. Grain boundaries are not visible. However, lamellar enrichment of dispersoids can be seen. Similar experiments were done using similar materials in.

Surface Effects

For diffusion welding, a very good quality of surfaces is a pre-requisite. Surfaces must be free of single deep scratches preventing vacuum-tight joints and of impurities from machining. Careful cleaning using surfactants and subsequent rinsing with ethanol or acetone are required. Gloves free of powder should be used for handling.

The number of stacked layers will also influence deformation at the given diffusion welding parameters, since multiple surfaces are approached and levelled. Hence, it is not possible to give a certain percentage of deformation to achieve highly vacuum-tight joints. Deformation also depends on the composition of the material.

Influence of Roughness

A pre-requisite for solid-state diffusion is a very good contact of the mating surfaces on the atomic level .The diffusion welding process can be divided into several phases. In the beginning, surfaces are approached by the deformation of asperities. At local spots, diffusion starts on the atomic level. Between these centres, pores remain which must be closed subsequently by volume diffusion. For this, the density of vacancies and, hence, the temperature and bonding time are essential. Additionally, temperature affects the grain growth and deformation.

Roughness influences the formation of a monolithic bond by the height and shape of asperities and the distance in between, forming temporary pores that must be filled. Perfectly smooth surfaces made, e.g., by diamond fly cutting may prevent local deformation because asperities are lacking. Shape and height of asperities, in conjunction with bearing pressure, define the local deformation behaviour.

Asperities may also help penetrate surface passivation layers, thus producing local initial metallic contact.

Passivation Layers

Some metals and alloys like aluminium, stainless steel, nickel-based alloys or titanium spontaneously form surface passivation layers. They consist mainly of oxides of the base metal; some alloying elements may be enriched. Often, oxygen is blocked to prevent further oxidation and the passivation layers are responsible for the good corrosion resistance in aqueous media or hot gases. Especially for aluminium, formation of passivation layers cannot be avoided completely. The thickness of these passivation layers is in the range of 2–20 nm depending on the type of metal and the content of alloying elements. Of course, composition, thickness and nature of passivation layers differ for normal austenitic stainless steel, heat-resistant steels or nickel-based alloys. Hence, the diffusion welding process must be optimised and the joint must be checked for grain growth across the bonding plane. High vacuum tightness is a necessary but not a sufficient criterion.

Figure: Diffusion-welded joint of Hastelloy C-22 (2.4602). T = 1100 °C, t = 70 min, p = 12 MPa ≥ high vacuum tightness. Subsequently, solution annealed at T = 1125 °C, t = 1 h, water-quenched ≥ leaky.

A certain roughness may help penetrate this layer by local deformation. Hence, the passivation layer comes into contact with matrix material. Passivation layers may be removed by chemical pickling. Even if subsequent formation of a new passivation layer may not be prevented, at least a reproducible surface condition is created.

Long bonding durations and high temperatures above 80% of the starting melting temperature should be preferred in this case.

Another approach is to remove the surface passivation layer, e.g., by sputtering with argon ions. Subsequently, a layer of a different metal may be deposited, which is not that susceptible to oxidation, e.g., gold or silver, and may temporarily form a low-melting alloy helping to create a bond.

For titanium, the passivation layer is soluble in the matrix material and diffusion welding of titanium is widely used, e.g., in the aerospace industry. In Figure, a very good bond between thin micro-structured layers can be seen. For parts consisting of multiple thin sheets, however, it has to be kept in mind that grades 1–4 differ slightly only in terms of the contents of nitrogen, oxygen and iron, while the mechanical properties are changed dramatically. Consequently, the properties of diffusion-welded parts may be changed for inappropriate ratios of surface layers to bulk material.

Figure: Diffusion weld of titanium, grade 2 at T = 850 °C, t = 4 h, p = 13 MPa.

Influence of Temperature, Bearing Pressure, Bonding Time and Design

Influence of Bonding Temperature

For diffusion welding, the joining temperature is normally set in the range of about 80% of the melting temperature for a pure metal or of the starting melting temperature for alloys. Temperature is calculated in Kelvin. Obviously, similar to the appropriate temperature for recrystallization, the temperature for alloys should exceed this level. For materials with surface passivation layers, temperature should be even higher and the time longer, which changes the whole process in terms of creep rate and appropriate bearing pressure. When comparing diffusion welding of, e.g., pure aluminium (Ts = 660 °C) and AlMg3 (Ts = 610–640 °C), the whole process has to be optimised. Otherwise, welding will fail due to excessive deformation.

The influence of temperature is strongly non-linear. Keeping in mind the dependence of vacancy density on temperature, an increase of about 20 K can double the diffusion coefficient and, hence, drastically increase the creep rate for a given bearing pressure.

Influence of Bearing Pressure

Bearing pressure is responsible for joining the mating surfaces. Influence of the bearing pressure is contrary to that of temperature. When increasing the bonding time from 1 to 4 h, the deformation is not proportional but increased by a factor of about 2.5. Obviously, a certain minimum bearing pressure is necessary to facilitate the deformation of local contact areas of the sample depending on the temperature applied.

Additionally, deformation under the given conditions strongly depends on the aspect ratio of the part and on the frictional cross-section between the part and the die applying the load. Large format parts of low thickness are difficult to weld with a reproducible deformation.

If the parts contain internal thin-walled micro-structures, the deformation behaviour may be affected by grain boundary sliding. For comparison to the part displayed in Figure, a similar part with a format of 40 × 30 mm, containing 12 layers of unstructured foils with a thickness of 3.6 mm and an overall height of 13.84 mm was welded under the same conditions (T = 850 °C, t = 4 h, p = 2 MPa). Deformation of the 12 layers was 1.3% only, while overall deformation was 0.35%, showing the influence of both micro-structures and aspect ratio.

The effective bonding area of a part should be distributed uniformly across the part. Otherwise irregular deformation or sink marks may occur. To prevent this, compensating areas may be helpful.

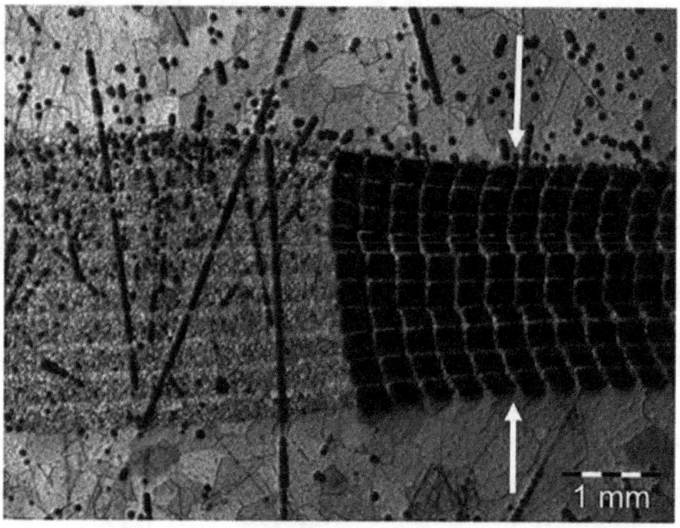

Figure: Irregular deformation on a part made of titanium with micro-structured sheets stacked in the same direction at T = 850 °C, t = 4 h, p = 10 MPa.

Influence of Bonding Time

Bonding time is required for conducting the diffusion process. After the initial step of approaching mating surfaces, time is needed to fill the pores left in between the local contact areas. Hence, a sufficient long bonding time is required.

Bonding time, together with temperature, affects deformation. As soon as creep takes place during diffusion bonding, a long bonding time makes it difficult to control deformation and design changes may have a major impact. Therefore, the diffusion welding process should be optimised for each serial application. It is difficult to weld prototypes of varying designs or materials without profound experience.

It is also hard to give a certain percentage of deformation to obtain a good diffusion bond. In fact, bonding quality depends on the number of layers to be bonded. In any case, the deformation behaviour depends not only on the composition of a material but also on its micro-structure.

Special Factors to be considered in Diffusion Welding of Micro-devices

Bearing pressure should be kept as low as possible, while, on the other hand, it must be sufficient to deform asperities and to increase the contact area during the bonding process.

The temperature should be sufficient for a high density of vacancies and for filling the pores by volume diffusion, which also depends on the bonding time. Micro-devices mainly consist of micro-structured multiple sheets. Channels may run in the same direction or cross-wisely, and the load-bearing structures may not proceed over the whole thickness for technical reasons.

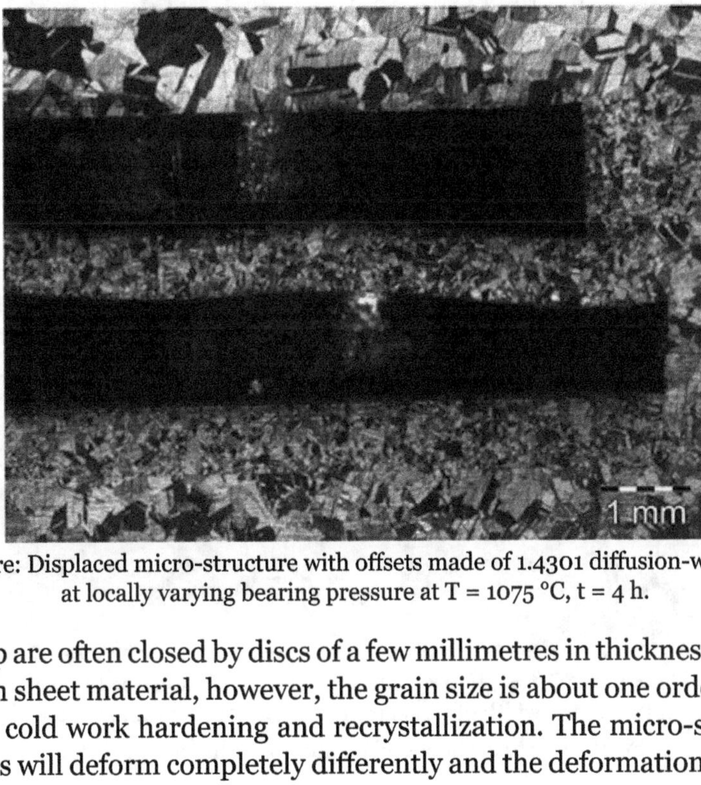

Figure: Displaced micro-structure with offsets made of 1.4301 diffusion-welded at locally varying bearing pressure at T = 1075 °C, t = 4 h.

Bottom and top are often closed by discs of a few millimetres in thickness, having coarse grains. For thin sheet material, however, the grain size is about one order of magnitude smaller due to cold work hardening and recrystallization. The micro-structured stack and thick plates will deform completely differently and the deformation will be concentrated mostly on the micro-structured section. An intelligent design may help achieve reasonable results.

Shapes of thin Walls in Micro-structures

The cross-section of thin walls may be important to the deformation behaviour: if the bearing pressure forces the material to creep, cross-sections of rectangular wall may be bent or deformed to a barrel shape. In the SEM of material with etched micro-channels on the right, however, the part is stabilised, since the bonding cross-section increases when deformation occurs.

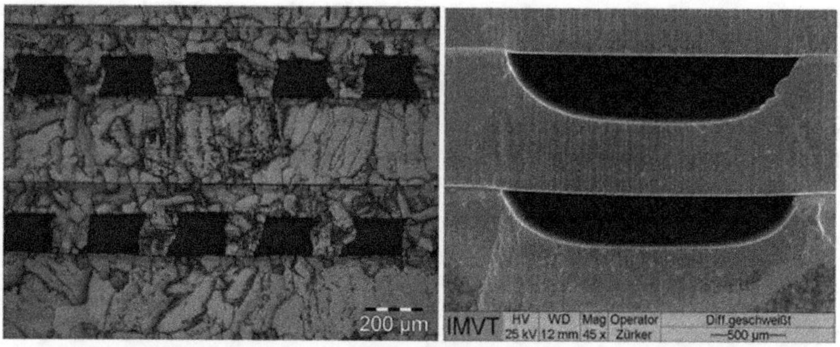

Figure: Impact of the cross-section of thin walls. Left: rectangular cross-section of thin walls. Right: self-reinforcing cross-section due to etched micro-channels.

The dimension of the walls should exceed the grain size of the material: in most cases, walls should be at least 100–200 µm in width. The aspect ratio should not exceed 1:1 for stability reasons, e.g., to avoid bending.

Impact of the Design of Mechanical Micro-structures, the Aspect Ratio and the Number of Layers on the Deformation

Moreover, the geometry of the micro-structured foils to be proper is important: the ratio between the thickness of the remaining bottom and the width of a trench should not exceed 1:1 to transmit sufficient bearing pressure to the next layer and to prevent lacking fusion.

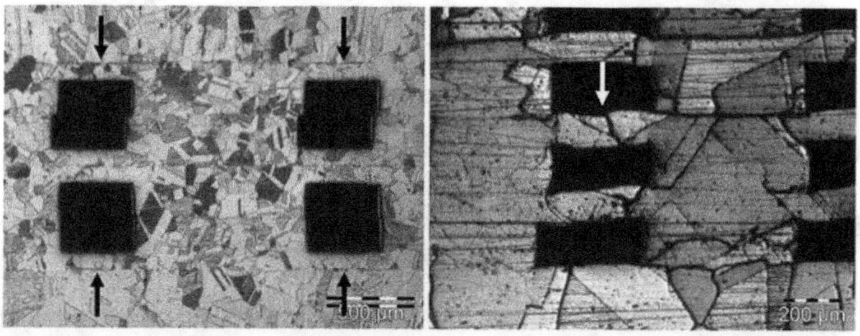

Figure: Left: 1.4301, T = 1075 °C, t = 4 h, p = 8 MPa, incomplete fusion due to insufficient thickness of the bottom in relation to the width of trenches. Right: 1.4876, T = 1250 °C, t = 1 h, p = 8 MPa, distortion due to grain boundary sliding within a thin bottom.

Depending on the application, a grain boundary crossing the remaining thickness of the bottom of a micro-channel should be avoided. In case of corrosion, this would be a favourable path for failure. During diffusion welding it causes local grain boundary sliding and distortion of the mechanical micro-structure.

Another topic is the aspect ratio of the parts to be welded, e.g., due to the different thermal expansion coefficients of the TZM-stamps, and the parts and deformation during the welding process, friction between both occurs. For a high aspect ratio in the range of one or more, a barrel-shaped profile results, accompanied by a high percentaged deformation. Flat parts, however, possess a low deformation at the same conditions.

For example, for disks of 160 mm in diameter, a deformation of 10% was obtained for a height of 10 mm for $T = 1075$ °C, $t = 4$ h, $p = 25$ MPa. For a height of 150 mm, however, the deformation was more than 33%.

Figure: Conical samples made of 1.4301, T = 1000 °C, t = 4 h, F = 17.55 kN, corresponding to 15–25 MPa. Left: Before diffusion welding. Middle: Five segments, deformation: 5.41 and 5.11%, respectively. Right: Sample made of 51 layers; deformation: 8.34%.

The number of layers affects the deformation obtained at the same conditions since the roughness of more surfaces must be levelled. For example, a conical sample consisting of 51 layers had a deformation more than 30% higher than the same sample geometry consisting of five segments only.

Equipment for Diffusion Welding

Diffusion bonding can be carried out using hot isostatic pressing (HIP) at a high isostatic pressure applied by argon of up to 2.500 bar or using a heated press with uniaxial load. For HIP, the parts must be placed inside a steel shield container which is evacuated before sealing. This makes the handling of the parts and the process itself rather expensive.

Additionally, sticking of the parts to the container must be prevented, e.g., by rock wool layers in between or boron nitride spray, or the container has to be machined off afterwards. When using fibrous materials, desorption from a high specific surface area at high temperatures has to be considered. However, also parts with an irregular bonding plane can be welded by HIP, since homogeneous pressure is applied. HIP is widespread and offered by service providers, e.g., ABRA Fluid AG.

Diffusion bonding using uniaxial heated presses is performed under a protective inert gas atmosphere or high vacuum. Only a few companies supply equipment for diffusion welding, e.g., PVA TePla AG, FormTech GmbH, TAV VACUUM FURNACES SPA and Centorr Vacuum Industries. Other companies such as MAYTEC Mess- und Regeltechnik GmbH and SYSTEC Vacuum Systems GmbH & Co.KG modify equipment like tensile testing machines or produce equipment for special needs. For this, a water-cooled vessel with a vacuum-sealed feed through for the dies is installed. The oven is heated indirectly by metallic heaters, and a vacuum in the order of 1E-05 Pa must be maintained for the protection of the heaters. Temperatures of not more than 1400 °C are sufficient for the most commonly used materials.

Figure: Diffusion bonding furnaces. Left: Maytec diffusion bonding furnace, maximum force 20 kN. Right: Systec diffusion bonding furnace, maximum force: 2 MN.

The stamps are often made of TZM, a molybdenum ODS-alloy, possessing still a high mechanical stiffness at high temperatures. However, the stability also depends on the thickness-to-diameter ratio and must be adapted to the forces transferred to the sample to prevent irregular deformation of the parts to be welded.

Due to the thermal mass of the equipment and to limit thermal stress, the heating rate and especially the cooling rate are low. PVA TePla AG also offers a rapid cooling technology for decreasing the cycle time.

During diffusion welding of stainless steel and nickel-based alloys under vacuum, chromium depletion takes place at the surface due to high partial pressure of chromium oxide. Hence, corrosion properties differ from a heat treatment in inert gas or air. For these materials, also enrichment of carbon must be prevented. Hence, unshielded heaters made of graphite are unsuitable.

Advantages and Disadvantages

Advantages

- The joint have same mechanical and physical properties as parent material.
- This process produces clean joint which is free from interface discontinuity and porosity.
- Both similar and dissimilar material can be joint by diffusion bonding process.
- It provides good dimension tolerance. So it is used to make precision components.
- Low running cost.

- It is simple in working.
- It does not use filler material, flux etc. which are used in arc welding process.
- It can weld complex shapes.

Disadvantages

- High initial or setup cost.
- It is time consuming process. It takes more time compare to other welding process.
- Surface preparations of welding plates are more critical and difficult.
- Size of the weld is limited according to equipment available.
- This process is not suitable for mass production.
- Highly depend on welding parameters like surface finish, welding material, temperature, pressure etc.

Application

- It is mostly used to weld refectory materials used in aerospace and nuclear industries.
- Diffusion bonding is used to weld titanium, zirconium and beryllium metals and its alloy.
- It can weld nickel alloy like Inconel, Wrought Udimet etc.
- It is used to weld dissimilar metals like Cu to Ti, Cu to Al etc.

References

- Solid-state-welding-process-principle-types-application-advantages-and-disadvantages: mech4study.com, Retrieved 21 July, 2019
- What-is-forge-welding-and-how-it-works: theweldingmaster.com, Retrieved 13 July, 2019
- Forge-welding-principle-working-application-advantages-and-disadvantages: mech4study.com, Retrieved 24 March, 2019
- Top-6-equipments-used-in-forge-welding-metallurgy, welding-science, metallurgy: yourarticlelibrary.com, Retrieved 1 June, 2019
- 6-steps-to-forge-weld-a-step-by-step-guide: blacksmithcode.com, Retrieved 21 March , 2019
- Friction-welding-operation-machines-and-applications, friction-welding, welding: yourarticlelibrary.com, Retrieved 12 April, 2019
- Top-5-solid-state-welding-processes-metallurgy, welding: yourarticlelibrary.com
- Diffusion-bonding-principle-working-application-advantages-and-disadvantages: mech4study.com, Retrieved 23 June, 2019

5

Welding Metallurgy and Weldability

Welding metallurgy involves the chemical, mechanical, and physical properties of metals. Weldability is the ability of a material to be welded. This chapter has been carefully written to provide an easy understanding of welding metallurgy and weldability of different materials.

Welding Metallurgy

Welding Metalurgy, in short, is the process of joining two or more metal components/parts/materials/joints/pipes/etc. Together to form a single whole piece/part/pipe.

Metallurgy is the science and study of metals, especially involving welding. Everyone should know basic metallurgy before they weld. When welding, you can alter the steel you are working on affecting it adversely. A good example is stainless which is alloyed with chromium to make it rust-resistant, and nickel to make it hard. If you burn the chromium out of the steel when welding, it will then rust. Stainless will also warp very easily from the heat if precautions are not taken.

Many people will put way too much weld on a joint thinking that more is better. Actually more can be worse because of the added heat. You are also wasting time and materials which wastes money. Studying metallurgy will help you know what amount of weld metal should be deposited.

Welding metallurgy involves the chemical, mechanical, and physical properties of metals.

- Chemical: One form of chemical metallurgy most everyone has heard of is rust. Rust is the oxidation of metal, or where oxygen gets into the metal and corrodes it. There is also corrosion where the atmosphere wastes away the metal. And in welding we are concerned with reduction, which is the removal of oxygen from the molten puddle.
- Mechanical: Mechanical metallurgy involves the way that metal acts under stresses and loads. Welding rods are numbered with the first 2 or 3 numbers telling the tensile strength of the rod per square inch of deposited metal. Tensile strength is the ability to resist being pulled apart. There are many different loads

and stresses involving brittleness, toughness, ductility, malleability, plasticity, shear, and others.

- Physical: Physical properties in metallurgy involve the metals being affected by the heat applied when welding. Grain size affects strength in metals, and grain size can be affected when welding. A good welder needs to know the thermal conductivity, melting point, and grain characteristics of the metal they are going to weld on.

One of the first things learned in metallurgy is the atomic structure of the metals. These structures consist of different "space lattices" which form different crystals. After you get a general idea of different structures in metals it will help you to understand the changes metals and steels go through when heated.

In studying metallurgy we find that carbon plays a big part in the strength of the steel we weld on. Steel is made harder by adding more carbon, but the more carbon added also makes it less ductile, thus more brittle. There are low, medium and high-carbon steels used in industry. The more carbon steel has, the more difficult to weld it will be.

Weldability

The American Welding Society defines weldability as, "the capacity of a metal to be welded under the fabrication conditions imposed into a specific, suitably designed structure and to perform satisfactorily in the intended service."

In simple terms, the ease with which metals can be fused without producing defects in the finished product is called "weldability". Many metals and thermoplastics can be welded, but some are easier to weld than others. A material's weldability is used to determine the welding process and to compare the final weld quality to other materials.

One of the main objectives in welding is to create a joint that is free of cracks and that can withstand the stresses placed on it. When nearly any welding process can be used on a metal and minimal effort is required to produce a sound weld, the metal is said to have "good weldability". In cases where a welder can choose from only a limited number of welding processes and must carefully prepare the joint and execute the welding procedure to create a strong weld, the metal is considered to have "poor weldability".

Weldability is considered as ease of accomplishing a satisfactory weld joint and can be in determined from quality of the weld joint, effort and cost required for developing the weld joint. Quality of the weld joint however, can be determined by many factors but the weld must fulfill the service requirements. The characteristics of the metal determining the quality of weld joint includes tendency to cracking, hardening and softening of heat affected zone (HAZ), oxidation, evaporation, structural modification and affinity to gases. While efforts required for producing sound weld joint are determined by the properties of metal system in consideration namely melting point, thermal expansion coefficient, thermal and electrical conductivity, defects inherent in base metal and surface condition. All the factors adversely affecting the weld quality and increasing the efforts (& skill required) for producing a satisfactory weld joint will in turn be decreasing the weldability of metal.

In view of above, it can be said that weldability of metal is not an intrinsic property as it is influenced by:

- All steps related with welding procedure.
- Purpose of the weld joints.
- Fabrication conditions etc.

Welding of a metal using one process may show poor weldability (like Aluminium welding with SMA welding process) and good weldability when the same metal is welded with some other welding process (Aluminium welding with TIG/MIG). Similarly, a steel weld joint may perform well under normal atmospheric conditions and the same may exhibit very poor toughness and ductility at very low temperature condition. Steps of the welding procedure namely preparation of surface and edge, preheating, welding process, welding parameters, post weld treatment such as relieving the residual stresses, can influence the weldability of metal appreciably. Therefore, weldability of a metal is considered as a relative term.

Factors Affecting Weldability

There are several factors which influence the weldability of metals. Below are some of the most important ones:

- Metallurgy: The science of heating or manipulating metals to produce desired properties or shapes in them.
- Welding Process: There are more than 67 welding processes. Various factors set them apart: how the heat and pressure are applied, how much heat and pressure are used, and the type of equipment employed.
- Joint Design: The combination of the dimensions required for the welded joint and the geometry of the joint.
- Weld Preparation: Weld preparation is a set of techniques to execute prior to welding to prevent defects in the weld. For example, one practice is to clean base metal before welding.
- Melting Point: The temperature that must be reached for a solid substance to melt or fuse. When a metal has a medium melting point, it has better weldability.

- Electrical Resistance: A metal's opposition, or resistance, to the flow of electrical current. Metals with a high electrical resistance require more heat energy to weld and this makes them have poor weldability.

The heating and cooling cycles inherent to most forms of welding can create strains and stresses in the weld. They also affect physical, chemical, and metallurgical changes in the metal. When such changes make a metal prone to poor weldability, adjustments can be made to improve the quality of the weld.

- Shielding Gas: Some types of metals, such as copper and aluminum, require a gas to shield them from atmospheric contaminants during welding. Selecting the correct shielding gas for the application in the right quantity can minimize the chance of weld defects.
- Welding Process: Welding charts are available to use as a reference for which process to choose for a metal.
- Filler Metal: Choosing the wrong filler metal can cause defects in the weld, such as cracks and porosity. A good general rule is to select a filler metal that is stronger than the base metal.
- Preheat and Postheat: Brittle metals are prone to cracking during welding. Heating the metal prior to welding and afterward can alleviate this problem.
- Welding Procedure: Weld quality can be dependent on the number of welds, their length, and the size of the weld bead. Several small welds can be more effective than a few large welds.

While all metals can be welded, some fuse more easily than others. However, an understanding of the factors that influence weldability and the practices to improve it can help a welder consistently produce strong and sound welds.

Imperfections in Welds

For steel there are few major failure modes or weld imperfections by which weldability can be measured:

Hydrogen-induced Cold Cracking

A characteristic feature of high carbon and low alloy steels is that the HAZ immediately adjacent to the weld hardens on welding with an attendant risk of cold (hydrogen) cracking. The weldability of steel, with regard to hydrogen-induced cold cracking, is inversely proportional to the hardenability of the steel, which measures the ease of forming martensite during heat treatment. The hardenability of steel depends on its chemical composition, with greater quantities of carbon and other alloying elements resulting in a higher hardenability and thus a lower weldability. In order to be able to judge alloys made up of many distinct materials, a measure known as the equivalent carbon content is used to compare the relative weldabilities of different alloys by comparing their properties to a plain carbon steel. The effect on weldability of elements like chromium and vanadium, while not as great as carbon, is more significant than that of copper and nickel, for example. As the equivalent carbon content rises, the weldability of the alloy decreases.

High-strength low-alloy steels (HSLA) were developed especially for welding applications during the 1970s, and these generally easy to weld materials have good strength, making them ideal for many welding applications.

Stainless steels, because of their high chromium content, tend to behave differently with respect to weldability than other steels. Austenitic grades of stainless steels tend to be the most weldable, but they are especially susceptible to distortion due to their high coefficient of thermal expansion. Some alloys of this type are prone to cracking and reduced corrosion resistance as well. Hot cracking is possible if the amount of ferrite in the weld is not controlled—to alleviate the problem, an electrode is used that deposits a weld metal containing a small amount of ferrite. Other types of stainless steels, such as ferritic and martensitic stainless steels, are not as easily welded, and must often be preheated and welded with special electrodes.

Lamellar Tearing

Lamellar tearing is a type of welding defect that occurs in rolled steel plates that have been welded together due to shrinkage forces perpendicular to the faces of the plates. Since the 1970s, changes in manufacturing practices limiting the amount of sulfur used have greatly reduced the incidence of this problem.

Lamellar tearing is caused mainly by sulfurous inclusions in the material. Other causes include an excess of hydrogen in the alloy. This defect can be mitigated by keeping the amount of sulfur in the steel alloy below 0.005%. Adding rare earth elements, zirconium, or calcium to the alloy to control the configuration of sulfur inclusions throughout the metal lattice can also mitigate the problem.

Modifying the construction process to use casted or forged parts in place of welded parts can eliminate this problem, as Lamellar tearing only occurs in welded parts.

Porosity

Porosity is formed by entrapment of discrete pockets of gas in the solidifying weld pool. The gas may originate from poor gas shielding, surface contaminants such as rust or grease, or insufficient deoxidants in the parent metal (autogenous weld), electrode or filler wire. A particularly severe form of porosity is ‘wormholes’, caused by gross surface contamination or welding with damp electrodes.

The presence of manganese and silicon in the parent metal, electrode and filler wire is beneficial as they act as deoxidants combining with entrapped air in the weld pool to form slag. Rimming steels with a high oxygen content, can only be welded satisfactorily with a consumable which adds aluminium to the weld pool.

To obtain sound porosity-free welds, the joint area should be cleaned and degreased before welding. Primer coatings should be removed unless considered suitable for welding by that particular process and procedure. When using gas shielded processes, the material surface demands more rigorous cleaning, such as by degreasing, grinding or machining, followed by final degreasing, and the arc must be protected from draughts.

Welding Metallurgy and Weldability of Steel

To understand the weldability of steel, it important to look into the different phases, phase mixtures and intermetallic generally found in steel besides the changes in phase that can occur during welding due to heating and cooling cycles. All these aspects can be understood by going through following section presenting significance of Fe-C diagram, time-temperature-transformation diagram and continuous cooling transformation diagram.

Fe-C Equilibrium Phase Diagram

Fe-C diagram is also called iron-iron carbide diagram because these are the two main constituents observed at the room temperature in steel while the presence of other phases depends on the type and amount of alloying elements. It shows the various phase transformations as a function of temperature on heating/cooling under equilibrium conditions.

Allotropy and Critical Temperatures

Change in crystal structure of an element with rise in temperature is termed as allotropy. Iron shows the allotropic behaviour at temperature 910 °C and 1390 °C. Iron changes its crystal structure first from BCC to FCC at 910 °C and then from FCC to BCC at 1390 °C. Therefore, solubility of carbon in iron varies with temperature especially above 910 °C and 1390 °C.

Isothermal Transformations in Fe-C Diagram

Three are main reaction points in Fe-C diagram namely peritectic, eutectoid and eutectic, which are of great academic and practical importance. All three reactions take place at a fixed temperature and composition.

Peritecitc reaction

$$\text{Liquid metal} + \delta\text{-iron (B. C. C.)} \xrightleftharpoons{1495\ ^{0}\text{C and } 0.18\%\ \text{C}} \gamma\text{-iron (F. C. C.)}$$

Eutectic reaction

$$\text{Liquid metal} \xrightleftharpoons{1130\ ^{0}\text{C and } 4.3\%\ \text{C}} \gamma\text{-iron (F. C. C.)} + Fe_3C$$

Eutectoid reaction

$$\text{Solid } (\gamma\text{-iron}) \xrightleftharpoons{727\ ^{0}\text{C and } 0.8\%\ \text{C}} \alpha\text{-iron (B. C. C.)} + Fe_3C$$

Proportions of various phases of these transformations can be obtained using lever rule. Fulcrum of the lever depends upon the alloy composition i.e. carbon content. Since these transformations take place at constant temperatures therefore it is easy to find the tie line. Terminal phases (and their compositions) can be obtained using the tie line and alloy compositions.

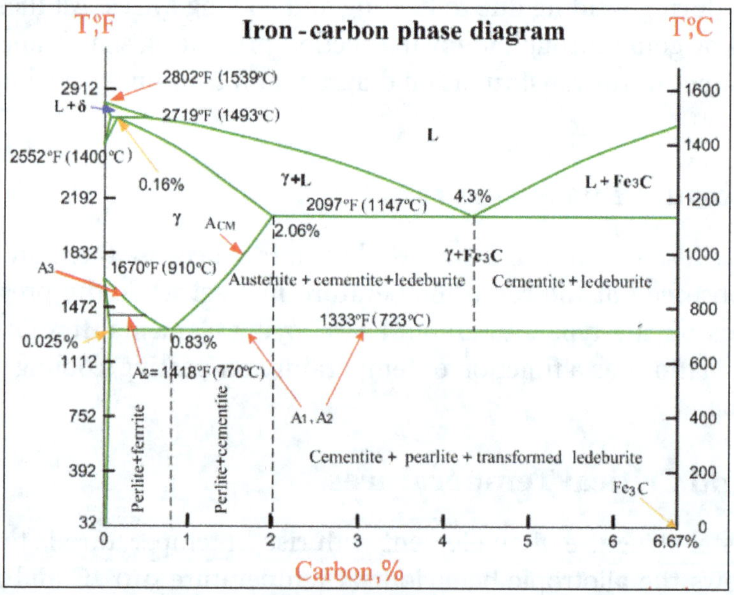

Figure: Fe-C diagram.

Phases appearing in Fe-C diagram are ferrite (α-Fe), austenite (γ-Fe), cementite (Fe_3C), δ iron and mixtures of phases such as pearlite (α-Fe + Fe_3C) and ledeburite (γ-Fe + Fe_3C). Pearlite and ledeburite are the result of eutectoid and eutectic reactions respectively.

Ferrite

Ferrite is an interstitial solid solution of carbon in iron having B. C. C. structure. Solubility of carbon in iron having B. C. C. structure at room temperature is about 0.005%

while that at eutectoid temperature (727 °C) is 0.025%. Ferrite is a soft, low strength, tough and ductile (50% elongation) phase.

Austenite

Austenite is an interstitial solid solution of carbon in iron having F. C. C. structure. This is not stable below eutectoid temperature (727 °C) in plain Fe-C system as it transforms into pearlite below this temperature. Solubility of carbon in iron having F. C. C. structure at temperature 13300 C is 2.0% while that at eutectoid temperature (727 °C) is 0.8%. Austenite is a comparatively harder, stronger, tougher but of lower less ductility (%elongation) than the ferrite.

Cementite

Cementite is an inter-metallic compound of iron and carbon i.e. iron carbide (Fe_3C). Cementite contains 6.67% of carbon and has orthorhombic structure. It is the hardest amongst all phases appearing in Fe-C diagram. Its hardness is extremely high while strength is very poor.

Pearlite

Pearliteis a phase mixture of ferrite and cementite and is a result of eutectoid transformation. The pearlite has alternate layers (lamellas) of cementite and ferrite. Strength of pearlite is more than any of the individual phases of which it is made. Mechanical properties i.e. strength, ductility, toughness and hardness of pearlite depends on the inter-lamellar spacing. Thinner plates (layers) of alpha ferrite, Fe_3C results in better mechanical properties.

Ledeburite

Ledeburite is also a phase mixture of austenite and cementite and is formed as a result of eutectic transformation but is observed only above the eutectoid temperature in Fe-C diagram as below this temperature austenite of ledeburitetransforms into pearlite.

Effect of Phases on Mechanical Properties

It is important to note that every phase or phase mixture has its own mechanical properties. Some of the phases are very soft (ferrite) and some are extremely hard (cementite). Therefore, variation in proportions/relative amounts of these phases will affect the mechanical properties of steel as a whole. Increase in carbon content in steel linearly increases proportion of pearlite but at the cost of ferrite. Since ferrite is of low strength, soft and ductile while pearlite is hard, strong and of poor ductility and toughness, hence increase in percentage of pearlite increases strength, hardness and reduces the ductility and toughness of steel as a whole. Cementite appears as an individual phase only above eutectoid composition (steel having carbon > 0.8%). It tends to form a network along the grain boundary of pearlite depending upon carbon content. Complete isolation of pearlite colonies with the cementite (as a result of continuous network of cementite) decreases tensile strength and ductility because mechanical properties of the alloy/steel to a large extent depend upon the properties of phase which is continuous in alloy. Increase in the carbon content above the eutectoid compositions (0.8% C) therefore reduces the strength and ductility because in hypereutectoid steel, network of cementite is formed along the grain boundaries of pearlite and cementite has very low tensile strength (3.0MPa).

Phase Transformation

Steel having 0.8% carbon is known as eutectoid steel. Steels with carbon lesser than 0.8% are known as hypoeutectoid and those having more than 0.8% and less than 2% carbon are called hypereutectoid steels. Fe-C systems having carbon more than 2% are called cast irons. Cast iron having 4.3% carbon is known as eutectic cast iron. Cast irons with carbon less than 4.3 % are known as hypoeutectic and those having more than 4.3% are called hypereutectic cast iron.

Steel

Eutectoid Steel

Transformation of eutectoid steel into various phases on cooling from molten state to the room temperature can be shown by drawing a vertical line through the composition point on E axis (called composition line). No phase transformation occurs until temperature of liquid metal goes down from point 'a' to 'b' on the liquidus. At the point 'b' solidification starts and first of all austenite (with 0.3%C) is formed. Further decrease in temperature results in formation of more and more proportion of austenite whose composition changes along the solidus whereas composition of liquid changes along the liquidus. It shows that carbon content in molten metal increases with the reduction in temperature until it goes down to the point 'c'. On completion of solidification, steel attains the austenitic state thereafter no phase transformation takes place until it attains eutectoid temperature (723 °C). Reduction in temperature decreases the solubility of carbon in austenite. Austenite is supersaturated with carbon as temperature goes below the eutectoid temperature hence austenite rejects the excess carbon leading to the nucleation of cementite along the grain boundary. As rejection of carbon from austenite continues, cementite layer grows toward the center of austenite grain and a very small zone which is depleted of carbon is formed on both the sides of the cementite plate. As concentration of carbon in carbon depleted zone reduces to such an extent when it is enough to dissolve in iron having B. C. C. structure at that temperature austenite transforms into ferrite. Thickness of the cementite or ferrite plates depends on the rate of diffusion and time available for the transformation to take place. As this transformation occurs by nucleation and growth mechanism which is based on diffusion of atoms. Slower the cooling rate, greater is the time available for the transformation and phases to grow hence thicker cementite and ferrite plates are produced. This process continues until whole austenite is transformed into pearlite.

Hypoeutectoid Steel

Transformation of 0.18% C steel during the cooling from molten state to the room temperature is shown by vertical line through the composition point. It is observed that no phase transformation takes place until temperature of liquid metal goes down from point 'd' to 'e' on the liquidus. At the point 'e' solidification starts and first of all δ iron (with 0.03% C) is formed. Further, decrease in temperature results in formation of more and more δ iron whose composition changes along the solidus whereas composition of liquid changes along the liquidus. Liquidus shows that carbon content in molten metal increases with the reduction in temperature. At the temperature of 1495 °C, liquid metal (with 0.5% C) and solid δ iron (0.1% C) the transforms into austenite by peritectic reaction. On completion of solidification, steel attains the complete austenitic state thereafter no phase transformation takes place until it attains temperature corresponding to the point 'g' (i.e.850 °C). At about 850 °C, austenite transforms into the ferrite. Further, transformation of austenite into ferrite continues with reduction in temperature until eutectoid temperature is attained. During this transformation, the composition of austenite changes continuously with increase in carbon content in austenite. As eutectoid temperature is attained carbon content in austenite rises to 0.8% i.e. eutectoid composition. At this temperature, austenite of eutectoid composition transforms into pearlite. Proeutectoid phase i.e. ferrite, is formed first from austenite in hypoeutectoid steel and subsequently austenite transforms into pearlite. Hence ferrite is found along the grain boundaries of pearlite. Continuity of ferrite network depends upon the carbon content in hypoeutectoid steel. Reduction in carbon content increases the continuity of ferrite network around the pearlite grains. Increase in carbon content reduces the amount of ferrite and increases the amount of pearlite.

Hypereutectoid Steel

Transformation of hypereutectoid steel is similar to that of hypoeutectoid steel except that proeutectoid phase is cementite instead of ferrite. In this case no phase transformation takes place until temperature of liquid metal goes down from point 'h' to 'I' on the liquidus. At the point 'i' solidification starts and first of all austenite (with 0.8%C) is formed. Further, decrease in temperature results in formation of more and more amount of austenite whose composition changes along the solidus whereas composition of liquid metal goes along the liquidus. It shows that carbon content in molten metal increases with the reduction in temperature until it goes down to the point 'j'. On completion of solidification, steel attains the austenitic state and thereafter no phase transformation takes place until it attains temperature corresponding to the point 'k' (1000 °C). At eutectoid temperature, austenite becomes supersaturated with carbon hence austenite rejects the excess carbon which leads to the nucleation of cementite along the grain boundary. Further, decrease in temperature results in formation of more and more cementite from the austenite whose composition changes along the kl. As eutectoid temperature is attained, carbon content in austenite decreases up to the eutectoid composition (0.8%C). At this temperature, austenite of eutectoid composition transforms into pearlite. Proeutectoid phase i.e. cementite is formed first from austenite in hypereutectoid steel subsequently austenite transforms pearlite. Hence, cementite is found along the grain boundaries of pearlite.

Time Temperature Transformation (TTT) Diagram

This diagram shows the transformation of meta-stable austenite phase at constant temperature into various phases as a function of time. Therefore, it is also known as isothermal transformation diagram. Transformation of austenite into various phases such as pearlite, bainite and martensite depends on the transformation temperature. Time needed to start the transformation of austenite into pearlite or bainite is called incubation period which is initially low at lower transformation temperature and then increases as transformation temperature increases. TTT Diagram for eutectoid C- steel has C shape due to variation in time needed to start and end the transformation of austenite at different transformation temperature This curve has a nose at about 550 °C. Transformation of austenite into pearlite takes place on exposure at any constant temperature above the nose. It is observed that higher the transformation temperature more is the time required for starting and completing the transformation. Transformation of austenite into pearlite or bainite occurs by nucleation and growth process. Hence, at high temperature high growth rate, low nucleation rate coupled with longer transformation time results in coarse pearlitic structure whereas at low transformation temperatures fine pearlitic structure is produced because of low growth rate, high nucleation rate and short transformation time. High transformation temperature lowers the strength and hardness of steel owing to the coarse pearlitic structure.

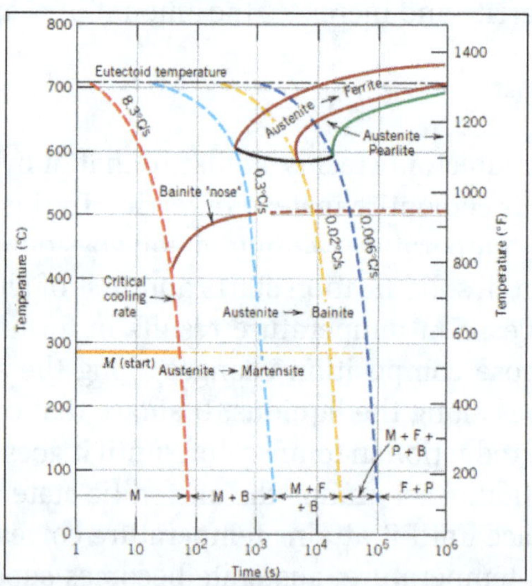

Figure: Time-temperature-transformation diagram:
TTT diagram superimposed with different cooling curves.

Transformation of austenite at a temperature below the nose of the curve results in bainitic structure. Bainite is a very fine intimate mixture of ferrite and cementite like pearlite. However, pearlite is a mixture of lamellar ferrite and cementite. That is why bainite offers much better strength, hardness and toughness than the pearlite. Degree of fineness of bainite also increases with the reduction in transformation temperature like pearlitic transformation. Bainite formed at high temperature is called feathery bainite while that formed at low temperature near the M_s temperature is called acicular bainite.

Transformation of austenite at a temperature below the M_s temperature results in hard and brittle phase called martensite structure. Martensite is a supersaturated solid solution of carbon in iron having body centered tetragonal (B.C.T) structure. This austenite to martensite transformation is athermal transformation as it takes place by diffusionless process. Rapid quenching/cooling from austenitic temperature to a temperature below the 'M_s' prevents any kind of atomic diffusion. Therefore carbon atoms, which are easily accommodated within FCC unit cell (in austenite) at high temperature, should be rejected at low temperature because of reduction in solid solubility. But at such a low temperature (below 'M_s') diffusion is prevented and that leads to formation of supersaturated solid solution of carbon in iron having BCC structure. This supersaturation of carbon in iron (BCC) causes the distortion of the BCC lattice structure and makes it BCT by increasing the c/a ratio more than 1. Degree of distortion is measured in terms of c/a ratio. This ratio depends up on the carbon content. Increase in carbon content up to 0.8%, increases the c/a ratio. The c/a ratio can be directly related with the increase in hardness, as there is linear relation between the two up to 0.8% carbon content.

Position of transformation lines and nose depends on steel composition, homogeneity of austenite and grain size. For each composition of steel, there will be just one TTT diagram. Steel other than eutectoid composition will have one more line initiating from the nose in TTT diagram corresponding to transformation of austenite into proeutectoid phase. In case of hypoeutectoid steel, first austenite forms ferrite as a proeutectoid phase, subsequently it transforms into pearlite whereas the proeutectoid phase for hypereutectoid steel is cementite.

Addition or reduction in carbon % in steel with respect to eutectoid composition shifts the nose of TTT diagram. For hypoeutectoid steels nose of the curve is shifted towards the left (reduced incubation time) whereas for hypereutectoid steels nose is shifted to right as compared to that for eutectoid steel. Temperature corresponding to start (M_s) and end (M_f) of martensite transformation is found a function of alloy compositions. Addition of alloying elements lowers these temperatures.

Continuous Cooling Transformation (C.C.T.) Diagram

CCT curve shows the transformation of austenite into various phases as a function of time at different cooling rates but not at constant temperature like in TTT diagram. This diagram is similar to that of TTT diagram except that under continuous cooling conditions (when temperature changes continuously) nose of the curve is shifted to right in downward direction and bainite transformation part as obtained in TTT diagram is absent in CCT diagram. Hence, continuous cooling diagram for eutectoid steel has only two lines above Ms, corresponding to the start and end of pearlite transformation. Various lines AB, KN, QR, ZO showing the reduction in temperature with time representing the different cooling rates called cooling curves. A very low cooling rate 'AB' results in the transformation of austenite into coarse pearlite starts at T_{s1} temperature and ends at T_{s2} temperature. Therefore, transformation takes place over a range of temperature from T_{s1} to T_{s2}. Such low cooling rates are used for annealing of steels which increases the softness. At somewhat higher cooling rate (KN) the transformation of austenite into pearlite starts at T_{s3} temperature and ends at T_{s4} temperature. Therefore, effective transformation takes place over a range of temperature from T_{s3} to T_{s4}. Grain size depends on the transformation temperature. High transformation temperature produces coarse grain. Since the continuous cooling conditions result in transformation over a range of temperature say T_{s1} to T_{s2} hence the grain size also varies accordingly. Therefore, at start of transformation coarser pearlite grains are formed than that at end of transformation. High cooling rate reduces the effective transformation temperature hence fine grain structure is produced. Such cooling rates are used for normalizing of steel. Normalizing increases the strength, hardness and toughness due to finer grain structure.

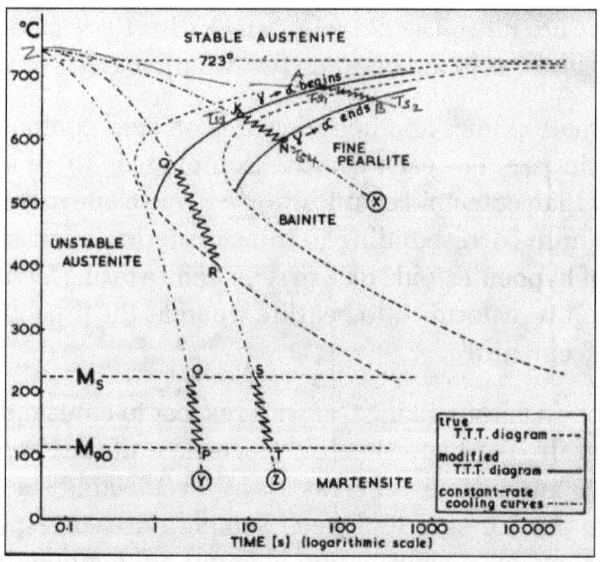

Figure: Continuous cooling transformation diagram.

Cooling curve 'QR' is tangential to 25% transformation line where 25% austenite has transformed into pearlite and 75% austenite is yet to transform. Thereafter, transformation of austenite stops and no further transformation takes place until its temperature goes down to M_s. As austenite crosses the M_s temperature remaining 75% austenite begins to transform into the martensite. This transformation line is expected to result in about 75% of martensite and approx. 25 % pearlite.

Further, higher cooling rate line 'ZO', which is tangential to the nose of the CCT diagram does not cause any transformation of austenite into pearlite. Austenite remains stable until M_s temperature is reached. Further reduction in temperature transforms the austenite into the martensite. Moreover, complete transformation of austenite into martensite depends on the quenching temperature. If quenching temperature is below the 'M_f' then only whole austenite is expected to transform into the martensite; otherwise some un-transformed austenite is left in steel called retained austenite. Retained austenite is comparatively soft; therefore, its presence reduces the hardness of steel. Amount of retained austenite depends on the quench temperature between M_s and M_f. Lower the transformation temperature (between M_s and M_f), smaller is the amount of retained austenite in steel. There is non-linear relationship between the amount of austenite transforming into martensite and quenching temperature in range of M_s and M_f. Minimum cooling rate that ensures complete transformation of austenite into martensite and avoids the formation of soft phases/phase mixtures (like pearlite) is called critical cooling rate (CCR).

Critical cooling rate depends on the position of the nose, which is governed by the alloy composition, grain size and homogeneity of austenite. To take into account the effect of all alloying elements on the critical cooling rate, carbon equivalent is used.

High carbon equivalent lowers the critical cooling rate hence less drastic cooling is required for hardening. In general, presence of all alloying elements (except Co) shifts the nose of CCT diagram towards right (conversely increases the incubation period to begin transformation) which in turn reduces the critical cooling rate and increases hardenability. Reduction in carbon content increases the critical cooling rate and makes hardening of steel more difficult.

Fine grained austenite starts the transformation earlier so nose of curves is shifted to left (reduces the incubation period for transformation). This increases the critical cooling rate. On the other hand coarse grain structure reduces the critical cooling rate which in turn increases the hardenability. Similarly inhomogeneous austenite (due incomplete transformation/austenitizing during heating) also reduces the transformation time, shifting nose of CCT curve to left and so increases the critical cooling rate.

Weldability of Steel and Composition

Weldability of steels can be judged by two parameters (a) cleanliness of weld metal and (b) properties of HAZ. Cleanliness of weld metal is related with presence of inclusion in the form of slag or gases whereas HAZ properties are primarily controlled by hardenability of the steel. Proper shielding of arc zone and degassing of molten weld metal can be used to control first factor. Proper shielding can be done by inactive gases released by combustion of electrode coatings in SMA or inert gases (Ar, He, Co_2) in case of TIG, MIG welding. Hardenability of steel is primarily governed by the composition. All the factors increasing the hardenability adversely affect the weldability because steel becomes more hard, brittle and sensitive to fracture/cracking, therefore it needs extra care. So, more the precautions should be taken to produce a sound weld joint.

Addition of all alloying elements (C, Mn, Ni, W, Cr etc.) except cobalt increases the hardenability which in turn decreases the weldability. To find the combined effect of alloying elements on hardenability/weldability, carbon equivalent (CE) is determined. The most of the carbon equivalent (CE) equations used to evaluate weldability depends type of steel i.e. alloy steel or carbon steel.

- Common CE equation for low alloy steel is as under:

  $CE = C + Mn/6 + (Cr + Mo + V)/5+(Ni + Cu)/15.$

  (elements are expressed in weight percent amounts)

- For low carbon steels and micro-alloy steels, CE is obtained using following equation:

  $CE = C + Si/25 + (Mn + Cr)/16 + (Cr + Ni + Mo)/20 + V/15.$

- From the Welding Journal, for low carbon, micro-alloyed steels, Ito-Besseyo carbon equivalent:

  $Ceq = C + Si/30 + (Mn + Cu + Cr)/20 + Ni/60 + Mo/15 + V/10 + 5*B.$

Since the effect of different alloying elements on hardenability of steel is different therefore, their influence on weldability will also be different. In general, high the CE steel needs high preheat temperature to produce crack free weld joint. Following point can be kept in mind as broad guidelines for welding steel:

$CE < 0.45$	No preheat required,
$0.45 < CE < 0.7$	$200-500^{\circ}C$ of preheat may be used
$CE > 0.7$	Cannot be welded.

Thickness of plate to be welded affects the cooling rate which in turn influences the hardening and cracking tendency. To take into account the thickness of plate above criteria is modified to get compensated carbon equivalent (CCE) relation.

$$CCE = CE + 0.00425\,t$$

where, t is the thickness of plate in mm,

$CE < 0.4$	No preheat required,
$0.45 < CE < 0.7$	$200-500\ ^{\circ}C$ of preheat may be used
$CE > 0.7$	Cannot be welded

From the weldability point of view, steels can be placed in five categories based on chemical composition, mechanical properties, heat treatment conditions, and high temperature properties: a) carbon steel, b) high strength low alloy steel, c) quench and tempered steel, d) heat treatable steel and e) Cr-Mo steel. These steels need to be welded in different forms such as sheets, plates, pipes, forgings etc. In case of steel welding, it is important to consider thickness of base metal as it affects the heat input, cooling rate and restraint conditions during welding.

Different Types of Steel and Welding

Carbon steel generally welded in as rolled condition (besides annealed and normalized one). The weldable carbon steel is mostly composed of carbon about 0.25 %, Mn up to 1.65%, Si up to 0.6% with residual amount of S and P below 0.05%. High strength low alloy steel (HSLA) is designed to have yield strength in range of 290-550 MPa using alloying concentration lesser than 1% in total. These can be welded in conditions same as that of carbon steel. Quench and tempered (Q & T) steels can be a carbon steel or HSLA steel category that are generally heat treated to impart yield strength in range of 350 to 1030 MPa. Heat treatable steels generally contain carbon more than carbon steel or HSLA steels, to increase their response to the heat treatment. However, presence of high carbon in these steels increases the hardenability which in turn decreases the weldability owing to increased embrittlement and cracking tendency of heat affected zone. Further, PWHT of heat treatable steel weld joints is done to enhance their toughness and induce ductility because presence of high carbon in these heat treatable steels. Cr-Mo steels are primarily designed to have high resistance to corrosion, thermal softening and creep at elevated temperature (up to 700 °C). Therefore, these are commonly used in petrochemical industries and thermal power plants. Weld joints of Cr-Mo steels are generally given PWHT to regain ductility, toughness, and corrosion resistance and reduce the residual stresses.

Common Problems in Steel Welding

Cracking of HAZ due to Hardening

The cooling rate experienced by the weld metaland HAZ during welding generally exceeds the critical cooling (CCR) which in turn increases the chances martensitic transformation. It is well known from the physical metallurgy of the steels that this transformation increases the hardness and brittleness and generates tensile residual stresses. This combination of high hardness and tensile residual stresses makes the steel prone to the cracking.

Cold Cracking

Another important effect of solid state transformation is the cold cracking. It is also termed as delayed/hydrogen induced cracking because these two factors (delay and hydrogen) are basically responsible for cold cracking. It can be observed that increase in stress decreases the time required for initiation and complete fracture by cold cracking.

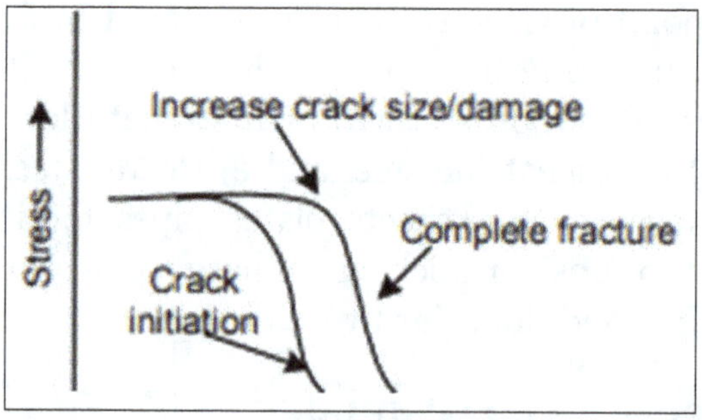

Figure: Schematic diagram showing stress vs. time relationship for fracture by cold crack.

Origin of this problem lies in the variation of solubility of hydrogen in the steel with the temperature. Reduction in temperature decreases solubility of hydrogen in solid state due to change in crystal structure from F. C. C. to B. C. C. High temperature transformation (like austenite to pearlite or bainite) allows escape of some of excess hydrogen (beyond the solubility) by diffusion. But in case of low temperature transformation (austenite into martensite), when rate of diffusion reduces significantly, hydrogen cannot escape and is trapped in steel as solid solution.

Hydrogen dissolved in atomic state at low temperature tends to diffuse out gradually toward the vacancies and other cavities. At these locations atomic hydrogen converts into diatomic H_2 gas and with time, continued diffusion of hydrogen towards these discontinuities as this gas starts to build up pressure in the cavities. If the pressure exceeds the fracture stress of metal, cavities expands by cracking. Cracking of metal increases the volume which in turn reduces the pressure. Due to continuous diffusion of hydrogen toward the cavities after some time again as pressure exceeds the fracture stress, and crack propagates further. This process of building up on pressure and propagation of cracks is repeated until compete fracture takes place without external load. Since this type cracking and fracture takes place after some time of welding hence it is called delayed cracking. Delay for complete fracture depends on the following factors:

- Hardenability of steel.
- Amount of hydrogen dissolved in atomic state.
- Magnitude of residual tensile stress.

Hardenability of steel affects the critical cooling rate. Steel of high hardenability promotes the martensitic transformation therefore it has high hardness and brittleness. High hardness increases the cracking tendency whereas soft and ductile metals reduce it. Crack tips are blunted in case of ductile metals so they reduce the cracking sensitivity and increases the stress level for fracture. As a result crack propagation rate is reduced in case of ductile and low strength metal. Therefore, steels of low hardenability will therefore minimize the cold/delayed cracking.

Larger the amount of dissolved hydrogen faster will be the delayed/hydrogen induced cracking.

Remedy:

- Use of low hydrogen electrodes.
- Preheating of plates to be welded.
- Use of austenitic electrodes.

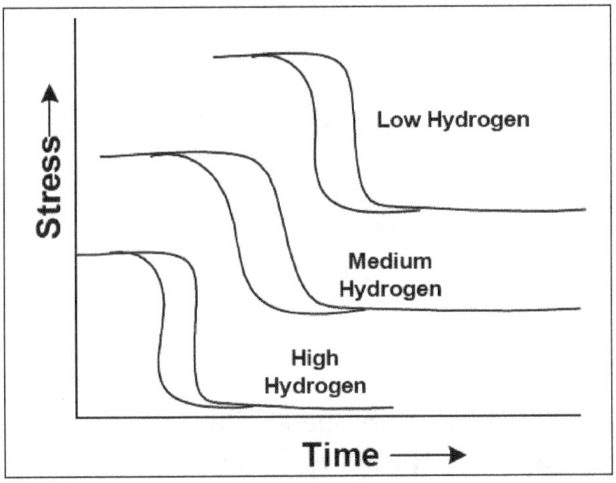

Figure: Schematic diagram showing effect of hydrogen concentration on cold cracking at different stress levels.

Use of low hydrogen electrodes will reduce the hydrogen content in weld metal. Preheating of the plate will reduce the cooing rate, which will allow longer time for gases to escape during the liquid to solid state and solid-solid transformation. It may also reduce the cooling rate below the critical cooling rate so that martensitic transformation can be avoided and austenite can be transformed into softer phases and phase mixtures like pearlite, bainite etc. These soft phases further reduce the cracking tendency. Use of austenitic electrode also avoids the martensite formation and provides mainly austenite matrix in weld zone. Austenite is a soft and tough phase having high solubility (%) for hydrogen. All these characteristics of austenite reduce the cold/delayed cracking.

Welding Metallurgy and Weldability of Stainless Steel

Stainless steel is not a single material, but a family of different types and grades, based on alloying chromium (Cr) with iron (Fe).

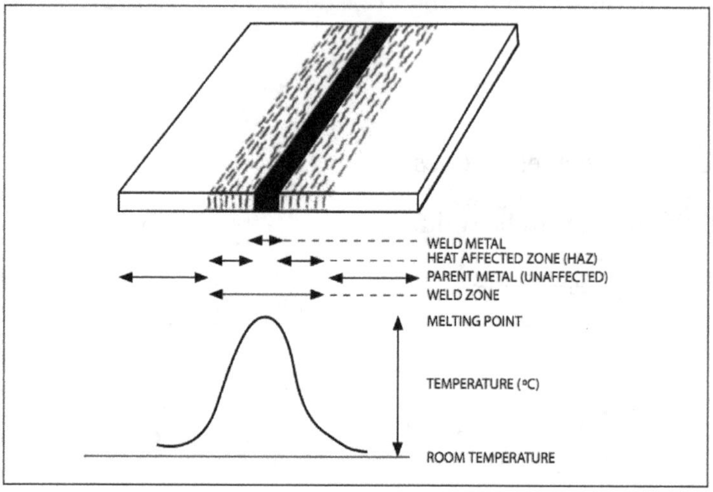

Figure: Schematic illustration of the three zones of welding and the related confined area of high temperature.

The Three Zones in Welding

In welding there are three zones of concern:

- The unaffected parent (or base) metal: The material being welded - at a relatively small distance from the weld it is only slightly warmed so that there is no change to its crystal structure, or to its properties.
- The heat affected zone (haz): The area of the parent metal adjacent to the weld metal is heated through a range of higher temperatures up to the melting point, which occurs at the junction with the weld metal.

  The actual temperature attained depends on the distance from the weld, the thermal conductivity of the steel and the amount of heat input.

  The high temperatures reached induce changes to the crystal structure of the parent metal, which in turn may affect its mechanical, physical and corrosion resisting properties.
- The solidified weld metal: The composition of this can be either that of the parent metal if no filter metal is used (known as autogenous welding) or that resulting from a mixture of filter and parent metal (depending on the extent to which the parent metal has been melted, thereby diluting the filler metal.

  Further, the solidified weld metal is a miniature "casting". Its cast grain structure will differ from the parent metal. This difference will be signification if the parent metal is plate or sheet, but also differ s, to a lesser extent, if the parent metal is a casting.

Weldability

The utilisation of welding in the fabrication of engineering components depends on the weldability of the steel. Weldability is not a precise or exactly quantifiable property, but rather implies the ability of the material to be joined by standard welding processes, so that the resultant mechanical, physical and chemical properties of the weld zone (i.e. both the weld metal and the HAZ) are at least equivalent to those of the parent metal.

Whereas all stainless steels may be welded, the weldability of the various classification of stainless steel varies to a large degree, due both to the different chemical compositions and crystal structures as well as to the resultant reactions of these to the thermal cycles of welding.

To a lesser extent the weldability of the different grades within any one classification can also vary.

Physical Properties of Stainless Steel

The physical properties of stainless steel may differ substantially from carbon and low-alloy steels (e.g. mild steels) and also between the various classifications of stainless steel.

The physical properties of major influence are:

- The coefficient of thermal expansion.
- The thermal conductivity.
- The electrical resistivity.
- The melting temperatures

Typical values of these physical properties for various types of steel are given in table.

Table: Typical physical properties for different types of steel.

Type of steel	Coefficient of thermal expansion (μm/m °c)	Thermal conductivity (w/m °c)	Electrical resistivity (μΩm)	Melting point (°c)
Plain chromium ferritic and martensitic stainless steel	11,3	24,0	0,060	1 430 to 1 510
Austenitic stainless steel	18,3	18,5	0,072	1 400 to 1 450
Utility ferritic stainless steel	12,5	25,0	0,057	1 430 to 1 510
Plain carbon and low-alloy steels	13,8	65,0	0,016	1 500 to 1520

Coefficient of Thermal Expansion

Austenitic stainless steel, in particular, will expand to a greater rate, than mild steel.

Allowance must be made for this expansion and contraction to control distortion and prevent high levels of thermal stress, particularly during cooling, which can cause weld zone cracking or high levels of residual stress.

Thermal Conductivity

Austenitic stainless steels, in particular, will conduct heat more slowly than mild steel.

This promotes sharp temperature gradients, which, together with the high thermal expansion, will cause that distortion to be confined to a small area. The weld zone will remain at high temperatures for a longer time, which is of critical importance in controlling interpass temperatures.

Electrical Resistance

The higher electrical resistance of stainless steel results in the generation of more heat for the same current.

This, together with the low heat conductivity, is an advantage when resistance welding processes are used. But, in other welding processes it can, if not controlled, result in too great a heat input. In manual metal arc (MMA) welding, the incorrect use of too high a welding current will cause overheating of the core wire of the electrode with associated damage to the flux coating.

Melting Temperatures

The lower melting temperatures of stainless steel compared with mild steel is an advantage, because less heat is required to produce fusion.

This in turn means faster welding for the same heat, or conversely less heat input for the same welding speed. These factors demonstrate that a change from welding mild steel to stainless steel should not be made in an undisciplined manner. With an appreciation of these differences, appropriate actions should be taken to avoid or minimise possible detrimental effects such as distortion and excessive heat input. These can include using smaller and more frequent tack welds, the use of copper back-up bars and chill plates or step welding.

Austenitic Stainless Steels

Their weldability is generally excellent, except for the free machining grades such as 303 or 303Se.

They may be readily welded by any of the standard welding processes such as manual metal arc (MMA), tungsten inert gas (TIG), metal active gas (MAG), submerged arc welding (SAW), resistance welding, friction welding, plasma arc, laser welding and electron beam welding.

Advances in welding technology (particularly those associated with developments in the field of welding consumables) and in steel making technology, have virtually eradicated the two major problems associated with the welding of austenitic stainless steel.

These are:

- Hot cracking within the weld zone:
    - Weld metal cracking.
    - HAZ cracking.
- Loss of corrosion resistance through sensitization.

Hot Cracking within the Weld Zone Weld Metal Cracking

Resistance across the weld and the crystal structure development in the weld metal are the major influencing factors. Fully austenitic weld metals are more prone to cracking.

Two forms of closely related hot cracking occur:

- Solidification Cracking takes place during and just after solidification. It is due to weld metal not being able to accommodate the thermal stresses during cooling. Mechanical failure takes place along the grain boundaries (intergranular cracking). This failure is more often than not the result of poor technique or poor consumable selection. Welding procedures suitable for mild steel are NOT directly applicable to stainless steel.
- Micro-fissuring occurs at temperatures very close to that of final solidification. It is intergranular in nature and is due to low melting point compounds segregating out at the grain boundaries. It is more difficult to avoid entirely and is not as dependent on welding conditions.

The presence of impurities, notably sulphur (S) and phosphorus (P) and to a slightly lesser extent silicon (Si) and niobium (Nb), has a marked influence on hot cracking. Modern steelmaking processes and impurity control have largely eliminated these problems.

Resistance to hot cracking is improved if the weld metal solidifies as two-phase austenite plus a small amount of retained ferrite.

This is due to three factors:

- The ferrite dissolves and retains in solution the impurities that would otherwise separate out.

- The ferrite present increases the grain boundary area available and thereby "dilutes" any grain boundary segregates that may occur.
- The ferrite fraction has a lower high temperature strength and therefore deforms without inducing cracking to accommodate the stresses.

  The chemical composition of most filler metals is therefore adjusted to result in a weld metal that contains 6-10% ferrite. This is indicated by the ferrite number (FN) normally given by the suppliers of welding consumables. (At low levels the FN may be taken to be equal to the ferrite percentage).

HAZ Cracking

This is also largely related to the impurity level. Ferrite, if present, has a beneficial effect. Fine grain sizes are less susceptible.

Wrought plate and sheet has a negligible ferrite content, but its fine grain size and low impurity levels are beneficial factors. Cast materials usually have a higher ferrite content that counteracts the normally associated higher impurity levels and coarser grain sizes.

Cracking in Perspective

An understanding of this has resulted in modifications to the composition of welding consumables and has virtually eliminated weld metal cracking (if a ferrite containing weld metal can be allowed).

Relatively few end-uses require ferrite-free, or very low ferrite content, weld deposits. The chemical industry sometimes considers the presence of ferrite an adverse factor because it can be prone to localised corrosion (e.g. in urea manufacture).

Cryogenic and continuous high temperature applications are two further such end-uses. Whereas generally there is a preference for the utilisation of ferrite containing weld metal, it must be stressed that significant amounts of fully austenitic (theoretically crackprone) weld metal are successfully deposited (e.g. 310, ferrite-free 316).

However, the necessary precautions should be taken in these instances to:

- Minimise stress/restraint.
- Use narrow weld/smaller stringer beads.

The above factors demonstrate the importance of the correct composition of the weld deposit.

Electrodes often have the necessary alloys content contained in the flux, and filler wires are specifically produced to close chemical composition limits. Therefore the following precautions should be noted:

- DO NOT use electrodes if the flux coating is damaged.
- DO NOT strip electrodes of the flux coating and use as filler wires.
- DO NOT use trimmings off sheet or random pieces of ordinary stainless steel wire as filler wire.

Loss of Corrosion Resistance through Sensitisation

Chromium has a high affinity for carbon at elevated temperatures. Heating the parent metal within the temperature of 450° - 850 °C, can cause Cr and C to combine as a complex chromium carbide, $Cr_{23}C_6$. The HAZ is unavoidably subjected to this temperature range during welding.

These carbides form preferentially on the grain boundaries, causing these areas to be depleted in Cr.

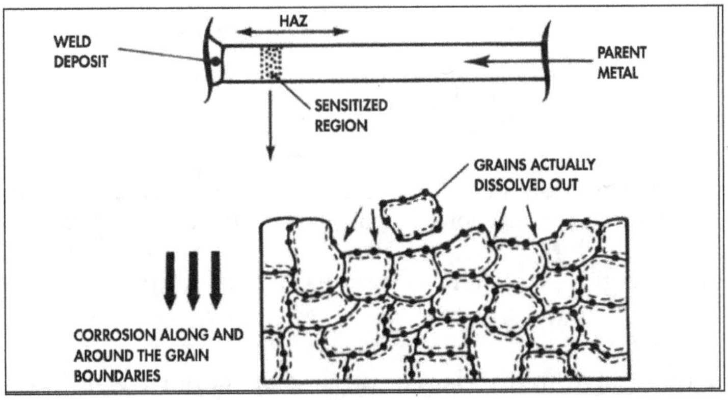

Figure: Schematic illustration of sensitisation within the HAZ; with associated chrome carbide and chromium depletion at the grain boundaries.

The steel has therefor become SENSITISED and is prone to localised corrosion occurring close to the weld deposit and along an intergranular path, due to the lower corrosion resistance of these Cr depleted grain boundary areas. This localised corrosion may only occur in specific conditions. This corrosion is termed WELD DECAY and is typified by a rough granular appearance of the corroded surface like grains of sugar.

Two methods are used to prevent weld decay:

- Stabilising alloying elements, such as titanium (Ti) and niobium (Nb), are added to the steel. These have a higher affinity for C than Cr does. Provided they are present in sufficient quantities, they preferentially combine with and lock up the carbon to form stable Ti or Nb carbides. Therefore, Cr carbides cannot form and the steel retains its corrosion resistance.

The typical stabilised grades of austenitic stainless steel are 321, 316 Ti (containing

Ti) and 347 (containing Nb). In the welding of these stabilised grades, it is generally recommended that a stabilised welding consumable be used. As Ti tends to be lost in the arc, Nb is used for stabilised consumables.

- The carbon content is reduced to very low levels. The carbon content of the steels is held to the very low level of 0,03% max C.

In addition to temperature, sensitisation is also dependent on carbon content and the time within the sensitisation temperature range. A 0,08% C steel can start to sensitise in under a minute, whereas a 0,03% C steel takes well over an hour. The typical low carbon grades ("L" grades) are 304L, 316L and 317L.

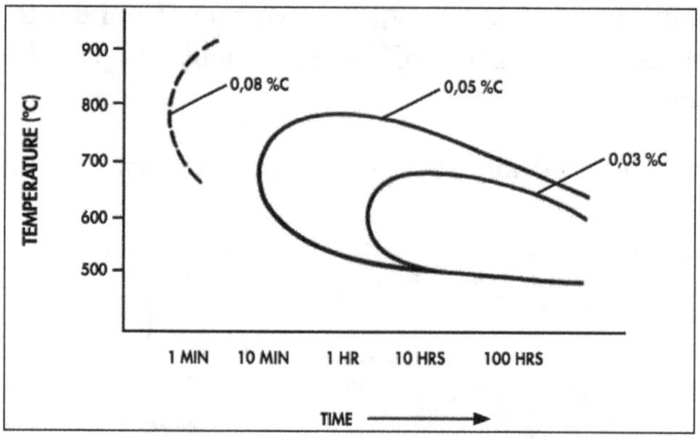

Figure: Schematic illustration of the major effect of carbon content on the time taken for sensitisation of a typical austenitic stainless steel.

In the welding of these grades either "L" grade, or Nb stabilised welding consumables must be used.

Due to either of the above methods, coupled with modern welding equipment that allows faster, lower heat input procedures, the classical form of weld decay through sensitisation, should not occur. Note that even in normal carbon grades of austenitic stainless steels, the carbon content is now controlled to lower levels than in the past, thus minimising the incidence of this problem.

However, responsibility rests with the welder to ensure that contamination of the weld from sources of carbon such as dust off the workshop floor, paint, grease and oil do not occur.

The presence of such carbon contamination is often overlooked, as the source is not always readily apparent (e.g. a mist of oil contained in the exhaust of air motor machines, handling electrodes with oily or greasy gloves).

Ferritic Stainless Steel

There are three sub-classifications of ferritic stainless steel to be considered:

- Standard ferritic stainless steel.

- Super ferritic stainless steel.
- Utility ferritic grades.

Standard Ferritic Stainless Steel

Standard ferritic stainless steel contained within the AISI 400 series. In general the weldability of these steels is poor and they suffer from inferior properties in the "as-welded" condition.

This is due to:

- Grain boundary breakdown at high temperatures (over 1 100 °C) forming larger, coarser grains.
- Formation of Cr carbide and Cr nitride precipitates and relatively high carbon martensite.

Typically these steels are brittle after welding and prone to corrosion within the HAZ. These deleterious properties generally render these steels unsuitable for engineering end-uses involving welded heavier gauge material.

Their normal applications as welded components are therefore limited to those utilising only thin gauge sheet material. Such thin material is usually welded using an autogenous welding process (TIG or resistance welding), although MAG welding is also used in high production processes.

If a filler metal is used, or if thicker gauges are welded, the austenitic filler metals, grades 308L and 309L, must be used. It is also necessary to minimise heat input and interpass temperatures.

Super Ferritic Stainless Steel

Several of the super-ferritic stainless steels available are proprietary grades but many are now included in international standards. Noticeable features of their chemical composition are:

- Increased Cr and additions of Mo to improve corrosion resistance.
- Very low C and N content.
- Stabilisation by additions of Ti, or Ti + Nb.

The weldability of these steels is better than that of standard ferritic steels.

Their corrosion resistance remains unimpaired and there is a far lower tendency for brittleness to develop. These improvements are chiefly related to the very low levels of C and N, plus the usual addition of stabilising elements. However, grain growth of the ferrite crystal structure can still detract from the properties in the "as-welded" condition.

Some manufacturers limit "good weldability" to a maximum thickness of 2,5mm but, in general, the weldability may be classed as "moderate" up to thicknesses of 5-6mm. They are best welded by resistance welding or autogenous TIG welding processes.

For welding thicker gauges use TIG or MAG and high nickel-based consumables. Matching filler metals tend to produce a coarse grained structure in the weld metal and are not recommended.

The inert gas welding processes are susceptible to CO_2, N, or H_2 impurities. Pure Argon is recommended for TIG and Argon +1 -2% O_2 for MIG. MMA welding is not recommended. Cleanliness is essential, particularly with respect to possible contamination by C. Heat input and interpass temperatures should be restricted to minimise grain growth.

Utility Ferritic Stainless Steel

These steels were developed to overcome the problems limiting the weldability of plain Cr ferritic stainless steels for engineering applications, but with sufficient Cr content to impart useful corrosion resistance.

If welded in accordance with recommended procedures the weldability is good in thicknesses up to 25mm to 30mm.

In the HAZ, the adverse effects of the thermal cycle of welding are minimised due to:

- Very low C and N content (which may be additionally stabilised by the addition of Ti).
- The mixed crystal structure of the steel at high temperatures.

The formation of chromium carbides and nitrides are reduced to a minimum, thereby overcoming both their embrittling effect and their > lowering of the corrosion resistance.

These steels develop a two-phase (i.e. mixed) crystal structure on heating to high temperatures. The presence of this mixed crystal structure strongly inhibits grain growth; any martensite formed is very low in carbon and is therefore tough.

On cooling, the crystal structure retains the mixed and small grained structure of the parent material. Some increase of hardness and strength occurs. The net result is that corrosion resistance is maintained and the properties are adequate for all applications, except severe engineering such as pressure vessels and highly stressed cyclic loading.

Thin gauge steels can be resistance welded. However, as most applications involve the use of thicker sheet and plate, MMA and MIG processes are most often used, TIG to a lesser extent. SAW is not recommended as the heat input is too high.

In the welding procedures it is of paramount importance to:

- Ensure cleanliness, especially prevent any carbon contamination.
- Minimise N pick-up by ensuring good arc shielding, especially in MIG and TIG processes.
- Ensure close control of heat input and minimise interpass temperatures. In this respect positional welding should be avoided if possible.
- If unavoidable, use multipass narrow stringer bead welds (smaller gauge electrodes with lower current) to minimise heat input.

For MMA and MAG welding, austenitic consumables are recommended.

The austenitic welding consumable gives an austenite/ferrite weld deposit, depending on the FN of the consumable and the degree of dilution.

Fabricator experience indicates a preference for grade 309L. It is of sufficiently high alloy content to cater for dilution by the parent metal, but not so highly alloyed to result in a fully austenitic weld metal that could be prone to hot cracking.

If utility ferritic steel is being welded to austenitic stainless steel, or to plain carbon (mild) or low alloy steel, an austenitic grade consumable must be used. The shielding gases used for MAG welding are either 98/argon 2% CO_2 mixtures. Pure CO_2 should not be used. If TIG welding is being carried out high purity argon should be used.

Martensitic Stainless Steel

Martensitic stainless steels are typified by those contained within the AISI 400 series. These steels have been specifically formulated to render them amenable to heat treatment by quenching and tempering to produce high strength and harness levels.

The thermal cycle of heating and rapid cooling that occurs in the welding process HAZ is equivalent to a quenching cycle. A high carbon martensite crystal structure is produced, which is highly stressed and extremely brittle. Therefore there is a strong tendency for cracking in the HAZ.

Martensitic stainless steel therefore possesses very poor weldability.

In general the applications that utilise martensitic stainless steel seldom involve the necessity for welding. However, if they have to be welded, the following courses of action may be adopted to limit failure by cracking:

- Preheat to between 200 °C and 300 °C. This prevents rapid cooling of the HAZ by surrounding parent metal. Interpass temperature should be at the same level.

- Weld the steels in the annealed condition and subsequently give the necessary heat treatment to produce the required properties.
- Use an austenitic consumable to improve ductility in the weld metal, but note that the mechanical properties in the weld metal will not match those of the parent metal. Matching consumables are available and if used, require extremely closely controlled conditions to prevent segregation and cracking of the weld metal.

Duplex Stainless Steel

Many of the duplex stainless steels available are proprietary grades developed and produced by different steel makers, but several are now included in international standards.

These steels consist of a two phase crystal structure of ferrite and austenite, The ferrite:austenite ratio is now generally controlled to about 50:50.

A wide range of new duplex grades containing manganese as a partial replacement for nickel are now known as "lean duplex grades" and the addition of nitrogen to increase both the stability of the austenite and improvements in mechanical properties is now common across all the duplex grades.

Control of the nitrogen content in the weld metal is now considered crucial and a variety of nitrogen containing shielding gases are now available to assist. As a result of continuous modification and development by the manufacturers the weldability of the duplex steels is now considered to be good, with engineering quality welds being achievable up to 50mm.

The weldability of these grades is determined by:

- The austenite changing to ferrite at high temperatures and subsequently reverting to austenite on cooling.
- The effect of the high temperatures on both the ferrite formed from the austenite and the ferrite fraction existing in the steel.

In welding the aim is to obtain conditions that maximise the reformation of austenite on cooling. The 50:50 ferrite:austenite crystal structure should be maintained after welding. Nitrogen (N) is an alloying element often used to attain this higher austenite content.

Duplex stainless steel has better heat conductivity than austenitic stainless steel. Faster cooling rates of the HAZ therefore occur. However, fast cooling retards the required reformation of austenite. The mechanical properties (particularly toughness) are critically dependant on achieving sufficiently high austenite content. Therefore, in comparison to the welding of other stainless steel, higher heat input and interpass temperatures are used when welding duplex stainless steels.

On the other hand, excessive heat inputs and slow cooling rates can promote an increase in the grain size of the ferrite and the formation of carbides that cause brittleness and loss of corrosion resistance.

A closely controlled balance between these two conditions has to be attained in the welding procedure. It is therefore recommended that the welding conditions for each grade and thickness of material be determined and the optimum be formulated in a qualified welding procedure, which thereafter should be closely controlled and monitored.

Because of the relatively lower coefficient of thermal expansion of duplex stainless steel, the stresses imposed in the weld zone are less than those experienced with austenitic stainless steel.

The welding processes that may be used include MMA, MIG/MAG, TIG and SAW.

Welding consumables for duplex stainless steel are essentially matching alloys, albeit often slightly modified in terms of chemical composition to ensure an optimum crystal structure of the weld metal. Ordinary austenitic stainless steel consumables are not suitable. Shielding gases used for MIG/ MAG and TIG processes are 98% HP Argon 2% N_2, heat input and interpass temperatures must be carefully controlled during the welding process.

Precipitation Hardening Stainless Steel

There are three sub-classifications of precipitation hardening stainless steel:

- Martensitic
- Semi-austenitic
- Austenitic

These are complex alloys and the welding and weldability of these is an extremely complicated subject.

These stainless steels can be successfully welded. However, the welding procedures are, in general, extremely complex and should not be entered into lightly without obtaining full details from the manufacturer of the specific alloy, training of the welders to be employed and using fully qualified welding procedures.

Pickling and Passivation

After welding of any of the types of stainless steel covered above, the heat affected zone will show some degree of oxide discolouration due to exposure to elevated temperatures. This oxide is not the same as the passive layer and will actually impair the corrosion resistance of the underlying stainless steel. It must therefore be removed before the fabrication enters service.

This can be done either by mechanical means (grinding, bead/grit blasting or polishing the affected area) or by pickling with suitable acid solutions or pastes, or a combination of both. The acid mixture generally used for pickling of stainless steel is a combination of nitric and hydrofluoric acids, normally applied hot. Details of the concentration, time and temperature will vary depending on the grade of stainless steel used and the degree of oxide present and assistance from a welding supplier, stainless steel cleaning company should be sought.

It is generally good practice to wash down the completed fabrication with a dilute solution of nitric acid in a mild detergent, followed by thorough rinsing in clean water. This will remove any dirt and grease, any mild steel contamination and will restore the passivity of the stainless steel to its optimum level.

Welding Metallurgy and Weldability of Aluminium Alloys

Aluminium and its alloys are widely used in welding industries due to economic advantages such as light weight, good corrosion resistance, high toughness, extreme temperature capabilities and easy recyclability. Aluminium alloys are used for construction of airplanes, cars, rail coaches and marine transports. Aluminium alloys are used in manufacture of tanks and pressure vessels because of their high specific strength, good heat conductivity and beneficial properties at low temperatures. Aluminium is the second most used metal after iron and steel in the industry; for example, aluminium is the second most used material taking about 15% of total body weight of average cars and about 34% in Audi A2. Welding is a means of joining metals by creating coalescence due to heat. The work piece is melted at the joint point (weld pool) that solidifies on cooling. Welding of aluminium alloys is important for fabricating structural constructions and mechanical fabrications like aircrafts. However, welding has problems and can be challenging. Welding defects common to aluminium include porosity, hot cracking and incomplete fusion.

Welding aluminium demands greater caution compared with steel, particularly as regards the amount of heat input and pre-weld cleaning, and that acceptable weld processes for aluminium joints are limited because the weldable thickness varies considerably with the different welding processes. It is therefore of interest to study the limitations facing aluminium welding, particularly joint- and process-specific limitations.

Surface-related Welding Considerations

A clean, smooth and protected surface is important in pre-weld aluminium structures to ensure good aluminium weldments except in high energy density welding processes like hybrid laser beam welding (LBW) (using pulsed metal inert gas (MIG)). It is therefore important to understand different surface-related phenomena and their effect on the weldability of the work piece. In addition, knowledge of preventative measures ensuring the attainment of acceptable welds, despite any adverse surface effects, is also important.

Presence of Aluminium Oxide Surface

Oxide formation in aluminium occurs due to the strong chemical affinity of aluminium for oxygen on exposure to air. The aluminium oxide thickness increases as a result of thermal treatment, moist storage conditions and electrochemical treatment (anodising). It is also important to note that Al_2O_3 melts at about 205 °C, while aluminium alloys melts at about 660 °C. Therefore, the layer is removed by pickling or dry machining just before weld. However, the difference in melting point is not a problem during the processing by means of high energy density welding processes; it can also be an advantage, for example, the presence of oxide layer during laser welding increases the absorptivity of aluminium and its alloys to laser radiation. It should be noted, that a main challenge in applying most joining technologies to aluminium is its tendency to form a thick, coherent oxide layer. This oxide layer has a melting temperature much higher than that of aluminium itself; moreover, it has a significant mechanical strength. Therefore, this oxide layer can remain as a solid film (or fractured in small particles) due to the flow of the molten material, even when the surrounding metal is molten. This can result in severe incomplete fusion defects. Therefore, the removal of the oxide layer just before welding is important.

The aluminium oxide layer is, furthermore, an electrical insulator, and the layer may sometimes be thick enough to prevent arc initiation. In MIG processes, a thick oxide layer can produce erratic electrical commutation in the gun's contact tube, resulting in poor welds.

It is thus evident that aluminium oxide has to be removed before welding because it compromises the quality of the weld. Generally, the oxide removal can be done by mechanical processes like brushing with a stainless steel brush, cutting with a saw or grinding with semi-flexible aluminium oxide grinding discs. Some welding processes enhance additional oxide removal processes, for example, in ultrasound metal welding processes (UW), oxides and contaminates are removed by high-frequency motion, thus providing metal–metal contact and allowing for the work pieces to bond properly. In hybrid laser MIG-welding of aluminium alloys, the MIG-welding process has a cleaning effect that removes the aluminium oxide layer. However, it is recommended that pre-weld cleaning of the weld surface should be carried out preferably by pickling or dry machining. In gas-shielded arc welding, aluminium oxide removal from the weld pool can be done by cathode etching (which is controlled chemical surface corrosion done to reveal the details of the microstructure). A direct current passes through the electrode connected to the positive pole of the power source. There is thus a flow of electrons from the work piece to the electrode and the ions flow in the opposite direction, bombarding the work piece surface. The aluminium oxide film is broken and dispersed by the ion bombardment, thereby allowing the flowing weld metal to fuse with the parent metal. It is advantageous to remove the aluminium oxide layer before welding because:

- It significantly reduces the amount of hydrogen porosity in the weld.
- It helps to improve the stability of the weld process especially in tungsten inert gas welding (TIG).
- It allows for complete fusion of the weld. Cathode cleaning is important in TIG process as the oxide starts to form immediately after wire brushing.

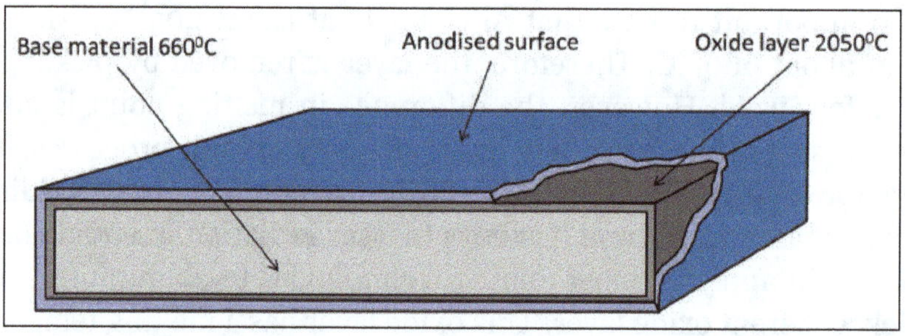

Figure: Schematic of aluminium showing its oxide layer and the anodised surface.

Table: Chemical treatments for cleaning and oxide removal.

Solution	Concentration	Temp (°C)	Procedure	Container material	Purpose
Nitric acid	50% water 50% HNO_3 (technical grade)	18–24	Immerse 15 min Rinse in cold water Rinse in hot water Dry	Stainless steel	Removal of thin oxide film for fusion welding
Sodium hydroxide followed by nitric acid	5% NaOH in water Concentrated HNO_3	70 18–24	Immerse for 10–60 s Rinse in cold water Immerse 30 s Rinse in cold water Rinse in hot water Dry	Mild steel Stainless steel	Removal of thick oxide film for all welding and brazing operations
Sulphuric chromic acid	5LH2SO_4 1.4 kg CrO_3 40 L water	70–80	Dip for 2–3 min Rinse in cold water Rinse in hot water Dry	Antimonial lead lined steel tank	Removal of films and stains from heat treating and oxide coatings
Phosphoric chromic acid	1.98 L of 75% H_3PO_3 0.65 kg of CrO_3 45 L of water	95	Dip for 5–10 min Rinse in cold water Rinse in hot water Dry	Stainless steel	Removal of anodic coatings

Aluminium oxide can also be removed by chemical etching or pickling. Table presents chemical treatments for oxide layer removal.

One of the causes of the oxide layer is from anodisation, which is an electrochemical process by which a metal surface is converted into a decorative, durable, corrosion resistant anodic oxide finish. Anodisation utilises the unique ability of amorphous alumina to build up an even porous morphology formed in alkaline and acidic electrolytes. During anodising, aluminium oxide is not applied like paint or plating. Rather, it is integrated fully with the underlying aluminium substrate. Therefore, it cannot peel or chip off. The anodic oxide structure is highly ordered and porous, thereby allowing for further processing like sealing and colouring.

The reasons for the utilisation of anodisation are to increase corrosion resistance and ensure the metal surface is fade proof for up to 50 years, to improve decorative appearance, to increase abrasion resistance and paint adhesion, to improve adhesive bonding and lubricity, to provide unique decorative colours or electrical insulation, to permit subsequent plating, to detect surface flaws, to increase emissivity and to permit application of photographic and lithographic emulsions.

Anodising of aluminium alloys is generally advantageous. However, it poses challenges for aluminium welding because the arc cleaning effect of the AC current cannot remove the double layer. Before welding, the anodised surface needs to be removed.

Shielding Gas Selection

Shielding gas protects the molten weld pool from the atmosphere, which is important because aluminium has a tendency to react with atmospheric air to form oxide and nitrides. The shielding gases commonly used in welding aluminium and its alloys are inert gases such as argon and helium.

Argon is used as a shielding gas for manual and automatic welding. Argon is cheaper than helium, and the use of argon produces a more stable arc and smoother welds. However, argon gives lower heat input and lower attainable welding speed, and therefore there is the possibility of a lack of fusion and porosity in thick sections. In addition, use of argon can result in a black sooty deposit on weld surfaces, although this can be wire brushed away. It has been observed that with helium shielding gas, the arc voltage is increased by 20%, resulting in a higher, hotter arc, deeper penetration and wider weld beads. This implies that the criticality of arc positioning (aids avoidance of missed edge and insufficient penetration defects) is lower with helium. There is a reduction in the level of porosity when helium shielding gas is used because the weld pool is hotter and there is slower cooling, which allows hydrogen to diffuse from the weld pool. Due to the higher heat produced, the use of helium allows that welding speeds up to three times higher than with argon. The high cost of helium and the inherent arc instability mean, however, that helium is used mainly in mechanised and automatic welding processes.

Table: MIG shielding gases for aluminium.

Metal transfer mode	Shielding gas	Characteristics
Spray transfer	100% Argon 35% Argon–65% Helium	Best metal transfer and arc stability, least spatter, and good cleaning action. Higher heat input than 100% argon; improved fusion characteristics on thicker material; minimises porosity.
Short circuiting	25% Argon–75% Helium Argon or Argon + Helium	Highest heat input; minimises porosity; least cleaning action Argon satisfactory on sheet metal; argon–helium preferred for thicker base material.

Table: Effect of shielding gas on aluminium welding.

Shielding gas	Relative effect (100% argon as the reference)		
	100% Ar	Ar + He	100% He
Gas flow	Nominal	Higher	Highest
Arc voltage (MIG)	Nominal	Higher	Highest
Arc (MIG)	Nominal Stability	More unstable	Most unstable
Weld seam width and depth	Nominal width and depth	Higher width	Highest width
		Shorter depth	Shortest Depth
Weld seam appearance	Nominal smoothness	Smoother	Smoothest
Penetration	Nominal depth and roundness	Deeper and more round	Deepest and Most round
Welding speed	Nominal welding speed	Higher attainability	Highest Attainability
Lack of fusion	Nominal	Lower	Lowest
Porosity	Nominal	Lower	Lowest
Pre-heating	Nominal	Less needed	Least needed
Heat production	Nominal warmth	Warmer work piece	Warmest work piece
Cost of shielding gas	Nominal price	More expensive	Most Expensive

It is common practice to use a mixture of helium and argon as it provides a compromise on the advantages of each gas. Common combinations are 50% or 75% of helium in argon, which allow for better productivity by increasing the welding speed and provide a wider tolerance for acceptable welds. The purity of the shielding gas is of importance. At the torch, not at the cylinder regulator, a minimum purity requirement of 99.998% and low moisture levels of less than - 50 °C (less than 39 parts per million (ppm) H_2O) are expected. Generally, the shielding gas should be selected with the following considerations:

- The gas must be able to generate plasma and a stable arc mechanism and characteristics.
- It should provide smooth detachment of molten metal from the wire and fulfil the desired mode of metal transfer.
- It should protect the welding head (in the arc's immediate vicinity), molten pool and wire tip from oxidation.
- It should help to attain good penetration and good weld bead profile.
- It should not affect the welding speed of the process.
- It should prevent undercutting tendencies.
- It should limit the need for post-weld cleaning.
- It should not be detrimental to the weld metal mechanical properties.

The recommended shielding gas for welding aluminium using pulsed MIG is argon (99.998%) at a flow rate of about 20 L/min. A mixture of argon and helium can also be used and even helium alone. Helium increases the weld penetration and offers higher arc energy and thus increased deposition rate, and it should be used when the section is greater than 50 mm.

Joint Types and Process Limitations

This topic considers eight industrially accepted welding processes and six joint types. Joint design is important because it costs money to buy weld metal. The fillet throat, weld accessibility and the functionality of the welded work piece are taken into consideration in this design. The six joints considered are butt, T-joint, corner, cruciform, edge and lap joint, which are derived from the three basic welding joints (fillet, lap and butt joints). Joint designs are based on the strength requirements, the alloys to be joined, the thickness of the material, the joint type and location, weld accessibility and the welding process. Before choosing the joint design, it is important to note that welding in the flat or downward position is preferable in all arc-welding processes, as there is the easier possibility of depositing high-quality weld metal at a high deposition rate in a flat position. Additionally, the weld pool is larger, allowing for a slower cooling and solidification rate, which enhances the escape of trapped gases in the weld pool. The flat position reduces weld porosity, reduces weld cost, and gives the best weld metal quality compared with other positions. The static tensile strength of the weld is determined by the throat thickness, which must be designed to ensure that it can carry the workload for which the weld is designed. Conventional TIG and MIG processes produce weld metal on the surface of a plate during bead-on-plate welds to a depth of 3 mm for TIG and 6 mm for MIG. Therefore, to attain complete penetration for welds over 3 mm (MIG) and 6 mm (TIG), there is the need for bevelling on butt joints, for example. The bevel can be single or double sided.

Welding Metallurgy and Weldability of Nickel Alloys

Nickel alloys can be joined reliably by all types of welding processes or methods, with the exception of forge welding and oxyacetylene welding. The wrought nickel alloys can be welded under conditions similar to those used to weld austenitic stainless steels. Cast nickel alloys, particularly those with a high silicon content, present difficulties in welding.

The most widely employed processes for welding the non-age-hardenable (solid-solution-strengthened) wrought nickel alloys are gas-tungsten arc welding (GTAW), gas-metal arc welding (GMAW), and shielded metal arc welding (SMAW). Submerged arc welding (SAW) and electroslag welding (ESW) have limited applicability, as does arc plasma welding (PAW). Although the GTAW process is preferred for welding the precipitation-hardenable alloys, both the GMAW and SMAW processes are also used.

Nickel alloys are usually welded in the solution-treated condition. Precipitation-hardenable (PH) alloys should be annealed before welding if they have undergone any operations that introduce high residual stresses.

Postweld Treatment: No postweld treatment, either thermal or chemical, is required to maintain or restore corrosion resistance, although in some cases a full solution anneal will improve corrosion resistance. Heat treatment may be necessary to meet specification requirements, such as stress relief of a fabricated structure to avoid age hardening or stress-corrosion cracking (SCC) of the weldment in hydrofluoric acid vapor or caustic soda. If welding induces moderate-to-high residual stresses, then the PH alloys would require a stress-relief anneal after welding and before aging.

Nickel and nickel alloys are susceptible to embrittlement by lead, sulfur, phosphorus, and other low-melting-point elements. These materials can exist in grease, oil, paint, marking crayons or inks, forming lubricants, cutting fluids, shop dirt, and processing chemicals.

Work-pieces must be completely free of foreign material before they are heated or welded. Shop dirt, oil and grease can be removed by either vapor degreasing or swabbing with acetone or another nontoxic solvent. Paint and other materials that are not soluble in degreasing solvents may require the use of methylene chloride, alkaline cleaners, or special proprietary compounds. If alkaline cleaners that contain sodium carbonate are used, then the cleaners themselves must be removed prior to welding. Spraying or scrubbing with hot water is recommended. Marking ink can usually be removed with alcohol.

Processing material that has become embedded in the work metal can be removed by grinding, abrasive blasting, and swabbing with 10% HCl solution, followed by a thorough water wash. Oxides must also be removed from the area involved in the welding operation, primarily because of the difference between the oxide and base metal melting points. Oxides are normally removed by grinding, machining, abrasive blasting or pickling.

Nickel alloys, both cast and wrought and either solid-solution-strengthened or precipitation-hardenable, can be welded by the GTAW process. The addition of filler is usually recommended. Direct current electrode negative (DCEN) is recommended for both manual and machine welding.

Shielding Gas: Either argon or helium, or a mixture of the two, is used as a shielding gas for welding nickel and nickel alloys. Additions of oxygen, carbon dioxide, or nitrogen to argon gas will usually cause porosity or erosion of the electrode. Argon with small quantities of hydrogen (typically 5%) can be used and may help avoid porosity in pure nickel, as well as aid in reducing oxide formation during welding.

Welding of Precipitation Hardenable Alloys

The PH alloys require special welding procedures because of their susceptibility to cracking. Cracks can occur in the base-metal heat-affected zone (HAZ) upon aging or in service at temperatures above the aging temperature, as a result of residual welding stress and stress induced by precipitation.

Before welding these alloys, a full-solution anneal is usually performed. After welding, the appropriate aging heat treatment is performed. To further improve alloy properties, a full anneal after welding, followed by a postweld heat treatment, can be incorporated in the welding procedure:

- Preweld and Postweld Treatments: Any part that has been subjected to severe bending, drawing or other forming operations should be annealed before welding. If possible, heating should be done in a controlled atmosphere furnace to limit oxidation and minimize subsequent surface cleaning.
- General Welding Procedures: Precipitation-hardenable alloys are usually welded by the GTAW process, but SMAW and GMAW processes are also applicable. Heat input during the welding operations should be held to a moderately low level in order to obtain the highest possible joint efficiency and minimize the extent of the HAZ. For multiple-bead or multiple-layer welds, many narrow stringer beads should be used, rather than a few large, heavy beads. Any oxides that form during welding should be removed by abrasive blasting or grinding. If such films are not removed as they accumulate on multiple-pass welds, then they can become thick enough to inhibit weld fusion and produce unacceptable laminar type oxide stringers along the weld axis.

Welding of Cast Nickel Alloys

Cast nickel alloys can be joined by the GTAW, GMAW and SMAW processes. For optimum results, casting should be solution annealed before welding to relieve some of the casting stresses and provide some homogenization of the cast structure. Light peening of solidified metal after the first pass will relieve stresses and, thus, reduce cracking at the junction of the weld metal and the cast metal. The peening of the subsequent passes is of little, if any, benefit. Stress relieving after welding is also desirable.

Minimizing Weld Defects

The defects and metallurgical difficulties encountered in the arc welding of nickel include:

- Porosity,
- Susceptibility to high-temperature embrittlement by sulfur and other contaminants,
- Cracking in the weld bead, caused by high heat input and excessive welding speeds,
- Stress-corrosion cracking in service.

Porosity: Oxygen carbon dioxide, nitrogen, or hydrogen can cause porosity in welds. In the SMAW and SAW processes porosity can be minimized by using electrodes that contain deoxidizing or nitride forming elements, such as aluminum and titanium. These elements have a strong affinity for oxygen and nitrogen and form stable compounds with them. Presence of deoxidizers in either type of electrode serves to reduce porosity. In addition, porosity is much less likely to occur in chromium-bearing nickel alloys than in non-chromium-bearing alloys.

In the GMAW and GTAW processes, porosity can be avoided by preventing the access of air to the molten weld metal. Gas backing on the underside of the weld is sometimes used. In the GTAW process the use of argon with up to 10% H_2 as a shielding gas helps to prevent porosity. Bubbles of hydrogen that form in the weld pool gather the diffusing hydrogen. Too much hydrogen (>15%) in the shielding gas can result in the hydrogen porosity.

Cracking: Hot shortness of welds can result from contamination by sulfur, lead, phosphorus, cadmium, zinc, tin, silver, boron, bismuth, or any other low-melting-point elements, which form intergranular films and cause severe liquid-metal embrittlement at elevated temperatures. Many of these elements are found in soldering and brazing filler metals.

Hot cracking of the weld metal usually results from such contamination. Cracking in heat-affected zone is often caused by intergranular penetration of contaminants from the base-metal surface. Sulfur, which is present in most cutting oils used for machining, is a common cause of cracking in nickel alloys.

Weld metal cracking also can be caused by heat input that is too high, as a result of high welding current and low welding speed. Welding speeds have a large effect on the solidification pattern of the weld. High welding speeds create a tear-drop molten weld pool, which leads to uncompetitive grain solidification at the center of the weld. At the weld centerline, residual elements will collect and cause centerline hot cracking or lower transverse tensile properties.

In addition, cracking may result from undue restraint. When conditions of the high restraint are present, as in circumferential welds that are self-restraining, all bead surfaces should be slightly convex. Although convex beads are virtually immune to centerline splitting, concave beads are particularly susceptible to centerline cracking. In addition, excessive width-to-depth or depth-to-width ratios can result in cracking may be internal (that is subface cracking).

Stress Corrosion Cracking: Nickel and nickel alloys do not experience any metallurgical changes, either in the weld metal or in the HAZ that affects normal corrosion resistance. When the alloys are intended to contact substances such as concentrated caustic soda, fluorosilicates, and some mercury salts, however, the welds may need to be stress relieved to avoid stress corrosion cracking. Nickel alloys have good resistance to dilute alkali and chloride solutions. Because resistance to stress-corrosion cracking increases with nickel content, the stress relieving of welds in the high-nickel-content alloys is not usually needed.

Effect of slag on weld metal: Because fabricated nickel alloys are ordinarily used in high-temperature service and in aqueous corrosive environments, all slag should be removed from finished weldments. If slag is not removed in this type of application, then crevices and accelerated corrosion can result. Slag inclusions between weld beads reduce the strength of the weld. Fluorides in the slag can react with moisture or elements in the environment to create highly corrosive compounds.

Welding Metallurgy and Weldability of Copper Alloys

Copper and copper alloys offer a unique combination of material properties that makes them advantageous for many manufacturing environments. They are widely used because of their excellent electrical and thermal conductivities, outstanding resistance to corrosion, ease of fabrication, and good strength and fatigue resistance. Other useful characteristics include spark resistance, metal-to-metal wear resistance, low-permeability properties, and distinctive color.

Welding Processes

In manufacturing, copper is often joined by welding. The arc welding processes are of primary concern. Arc welding can be performed using shielded metal arc welding (SMAW), gas-tungsten arc welding (GTAW), gas-metal arc welding (GMAW), plasma arc welding (PAW), and submerged arc welding (SAW).

Arc Welding Processes: Copper and most copper alloys can be joined by arc welding. Welding processes that use gas shielding are generally preferred, although SMAW can be used for many noncritical applications.

Argon, helium, or mixtures of the two are used as shielding gases for GTAW, PAW, and GMAW. Generally, argon is used when manually welding material is less than 3 mm thick, has low thermal conductivity, or both. Helium or a mixture of 75% helium and 25% argon is recommended for machine welding of thin sections and for manual welding of thicker sections of alloys that have high thermal conductivity. Small amounts of nitrogen can be added to the argon shielding gas to increase the effective heat input.

Shielded metal arc welding can be used to weld a wide range of thickness of copper alloys. Covered electrodes for SMAW of copper alloys are available in standard sizes ranging from 2.4 to 4.8 mm.

1. Gas-Tungsten Arc Welding: Gas-tungsten arc welding is well suited for copper and copper alloys because of its intense arc, which produces an extremely high temperature at the joint and a narrow heat-affected zone (HAZ).

In welding copper and the more thermally conductive copper alloys, the intensity of the arc is important in completing fusion with minimum heating of the surrounding, highly conductive base metal. A narrow HAZ is particularly desirable in the welding of copper alloys that have been precipitation hardened.

Many of the standard tungsten or alloyed tungsten electrodes can be used in GTAW of copper and copper alloys. The selection factors normally considered for tungsten electrodes apply in general to the copper and copper alloys. Except for the specific classes of copper alloys, thoriated tungsten (usually EWTh-2) is preferred for its better performance, longer life, and greater resistance to contamination.

2. Gas-Metal Arc Welding: Gas-metal arc welding is used to join of the coppers and copper alloys for thickness less than 3 mm, while GMAW is preferred for section thickness above 3 mm and for the joining of aluminum bronzes, silicon bronzes and copper-nickel alloys.

3. Plasma Arc Welding: The welding of coppers and copper alloys using PAW is comparable to GTAW of these alloys. Argon, helium, or mixtures of the two are used for the welding of all alloys. Hydrogen gas should never be used when welding coppers.

Plasma arc welding has two distinct advantages over GTAW: (1) the tungsten is concealed and entirely shielded, which greatly reduces contamination of the electrode, particularly for alloys with low-boiling-temperature constituents such as brasses, bronzes, phosphor bronzes, and aluminum bronzes, and (2) the constructed arc plume gives rise to higher arc energies while minimizing the growth of the HAZ. As with GTAW, current pulsation and current ramping may also be used. Plasma arc welding equipment has been miniaturized for intricate work, known as microplasma welding.

Plasma arc welding of coppers and copper alloys may be performed either autogenously or with filler metal. Filler metal selection is identical to that outlined for GTAW. Automation and mechanization of this process is readily performed and is preferable to GTAW where contamination can restrict production efficiencies. Welding positions for PAW are identical to those for GTAW. However, the plasma keyhole mode has been evaluated for thicker sections in a vertical-up position. Generally, all information presented for GTAW is applicable to PAW.

4. Submerged Arc Welding: The welding of thick gage material, such as pipe formed from heavy plate, can be achieved by continuous metal-arc operation under a granular flux. Effective deoxidation and slag-metal reactions to form the required weld-metal composition are critical and the SAW process is still under development for copper-base materials, A variation on this, process can be used for weld cladding or hard facing. Commercially available fluxes should be used for the copper-nickel alloys.

Alloy Metallurgy and Weldability

Many common metals are alloyed with copper to produce the various copper alloys. The most common alloying elements are aluminum, nickel, silicon, tin, and zinc. Other elements and metals are alloyed in small quantities to improve certain material characteristics, such as corrosion resistance or machinability.

Copper and its alloys are divided into nine major groups. These major groups are:

- Coppers, which contain a minimum of 99.3% Cu.
- High-copper alloys, which contain up to 5% alloying elements.
- Copper-zinc alloys (brasses), which contain up to 40% Zn.
- Copper-tin alloys (phosphor bronzes), which contain up to 10% Sn and 0.2% P.
- Copper-aluminum alloys (aluminum bronzes), which contain up to 10% Al.
- Copper-silicon alloys (silicon bronzes), which contain up to 3% Si.
- Copper-nickel alloys, which contain up to 30% Ni.
- Copper-zinc-nickel alloys (nickel silvers), which contain up to 7% Zn and 18% Ni.
- Special alloys, which contain alloying elements to enhance a specific property or characteristic, for example machinability.

Many copper alloys have common names, such as oxygen-free copper (99.95% Cu min), beryllium copper (0.02 to 0.2% Be), Muntz metal (Cu40Zn), Naval brass (Cu-39.5Zn-0.75Sn), and commercial bronze (Cu-10Zn).

Many of the physical properties of copper alloys are important to the welding processes, including melting temperature, coefficient of thermal expansion, and electrical and thermal conductivity. Certain alloying elements greatly decrease the electrical and thermal conductivities of copper and copper alloys.

Several alloying elements have pronounced effects on the weldability of copper and copper alloys. Small amounts of volatile, toxic alloying elements are often present in copper and its alloys. As a result, the requirement of an effective ventilation system to protect the welder and the welding machine operator is more critical then when welding ferrous metals.

Zinc reduces the weldability of all brasses in relative proportion to the percent of zinc in the alloy. Zinc has a low boiling temperature, which results in the production of toxic vapors when welding copper-zinc alloys.

Tin increases the hot-crack susceptibility during welding when present in amounts from 1 to 10%. Tin, when compared with zinc, is far less volatile and toxic. During the welding tin may preferentially oxidize relative to copper. The results will be an oxide entrapment, which may reduce the strength of the weldment.

Beryllium, aluminum, and nickel form tenacious oxides that must be removed prior to welding. The formation of these oxides during the welding process must be prevented by shielding gas or by fluxing, in conjunction with the use of the appropriate welding current. The oxides of nickel interfere with arc welding less than those beryllium or aluminum. Consequently, the nickel silvers and copper-nickel alloys are less sensitive to the type of welding current used during the process. Beryllium containing alloys also produce toxic fumes during the welding.

Silicon has a beneficial effect on the weldability of copper-silicon alloys because of its deoxidizing and fluxing actions.

Oxygen can cause porosity and reduce the strength of welds made in certain copper alloys that do not contain sufficient quantities of phosphorus or other deoxidizers. Oxygen may be found as a free gas or as cuprous oxide. Most commonly welded copper alloys contain deoxidizing element, usually phosphorus, silicon, aluminum, iron, or manganese.

Iron and manganese do not significantly affect the weldability of the alloys that contain them. Iron is typically present in some special brasses, aluminum bronzes, and copper-nickel alloys in amounts of 1.4 to 3.5%. Manganese is commonly used in these same alloys, but at lower concentrations than iron.

Free-Machining Additives: Lead, selenium, tellurium and sulfur are added to copper alloys to improve machinability. Bismuth is beginning to be used for this purpose as well when lead-free alloys are desired. These minor alloying agents, while improving machinability, significantly affect the weldability of copper alloys by rendering the alloys hot-crack susceptible. The adverse effect on weldability begins to be evident at about 0.05% of the additive and is more severe with larger concentrations. Lead is the most harmful of the alloying agents with respect to hot-crack susceptibility.

Factors Affecting Weldability

Besides the alloying elements that comprise a specific copper alloy, several other factors affect weldability. These factors are the thermal conductivity of the alloy being welded, the shielding gas, the type of current used during welding, the joint design, the welding position, and the surface condition and cleanliness.

- Effect of Thermal Conductivity: The behavior of copper and copper alloys during welding is strongly influenced by the thermal conductivity of the alloy. When welding commercial coppers and lightly alloyed copper materials with high thermal conductivities, the type of current and shielding gas must be selected to provide maximum heat input to the joint. This high heat input counteracts the rapid head dissipation away from the localized weld zone. Depending on section thickness, preheating may be required for copper alloys with lower thermal conductivities. The interpass temperature should be the same as for preheating. Copper alloys are not post-weld head treated as frequently as steels, but some alloys may require controlled cooling rates to minimize residual stresses and hot shortness.
- Welding Position: Due to the highly fluid nature of copper and its alloys, the flat position is used whenever possible for welding. The horizontal position is used in some fillet welding of comer joints and T-joints.
- Precipitation-Hardenable Alloys: The most important precipitation-hardening reactions are obtained with beryllium, chromium, boron, nickel, silicon, and zirconium. Care must be taken when welding precipitation-hardenable copper alloys to avoid oxidation and incomplete fusion. Whenever possible, the components should be welded in the annealed condition, and then the weldment should be given a precipitation-hardening heat treatment.
- Hot Cracking: Copper alloys, such as copper-tin and copper-nickel, are susceptible to hot cracking at solidification temperatures. This characteristic is exhibited in all copper alloys with a wide liquidus-to-solidus temperature range. Severe shrinkage stresses produce interdendritic separation during metal solidification. Hot cracking can be minimized by reducing restraint during welding, preheating to slow the cooling rate and reduce the magnitude of welding stresses, and reducing the size of the root opening and increasing the size of the root pass.
- Porosity: Certain elements (for example, zinc, cadmium, and phosphorus) have low boiling points. Vaporization of these elements during welding may result in porosity. When welding copper alloys containing these elements, porosity can be minimized by higher weld speeds and a filler metal low in these elements.

- Surface Condition: Grease and oxide on work surfaces should be removed before welding. Wire brushing or bright dipping can be used. Miliscale on the surfaces of aluminum bronzes and silicon bronzes is removed for a distance from the weld region of at least 13 mm, usually by mechanical means. Grease, paint, crayon marks, shop dirt, and similar contaminants on copper-nickel alloys may cause embrittlement and should be removed before welding. Miliscale on copper-nickel alloys must be removed by grinding or pickling; wire brushing is not effective.

References

- Metallurgy: keenovens.com, Retrieved 3 August, 2019
- Weldability-welding: theprocesspiping.com, Retrieved 19 May, 2019
- 3-Information-series-the-welding-of-stainless-steel-material: generalprofiling.co.za, Retrieved 28 April, 2019
- Aluminium-alloys-welding-processes-Challenges-joint-types-and-process-selection: researchgate.net, Retrieved 8 August , 2019
- Welding-of-nickel-alloys: nickel-alloys.net, Retrieved 9 May, 2019

6

Other Approaches to Welding and Joining

Many other approaches come under the domain of welding and joining which can be named as brazing, soldering and adhesive bonding. This chapter closely examines these approaches of welding and joining as well as welding process for plastics to provide an extensive understanding of the subject.

Brazing

Brazing is a versatile joining process. It is used to join most metals and alloys commonly used in engineering. It is a thermal joining process in which a molten brazing filler metal is drawn into a capillary gap between the metals being joined.

Brazing filler metals have a melting point above 450 °C but below the melting point of the metals being joined.

Soldering differs from brazing in that it takes place at temperatures below 450 °C, although both processes occur below the melting point of the metals being joined.

Properties of a Successfully Brazed Joint

- Strong and ductile: Well-made brazed joints can be at least as strong as the parent metals being joined and will withstand demanding service conditions.

- Leak tight: Brazing produces leak tight joints widely used on liquid and gas pipe-work installations.
- Electrical conductivity: Brazed joints offer good electrical conductivity and are used in applications where this property is important.
- Joint appearance: Brazed joints have smooth, neat fillets.

Terms Associated with the Brazing Process and Brazing Filler Metals

- Brazing alloy/filler metal: A filler metal which melts above 450 °C but below the metals being joined.
- Flux: A chemical compound applied to the parent metals to protect them from oxide formation during heating and to promote flow and wetting of the brazing filler metal.
- Parent metals: The materials to be joined by brazing.
- Wetting: When molten brazing alloy flows over and adheres to the parent metals.
- Joint gaps: The gap between the parent metals to be joined by the brazing filler metal.
- Capillary attraction: The force that draws a molten brazing alloy into the joint gap.
- Solidus: The highest temperature at which a filler metal is completely solid.
- Liquidus: The lowest temperature at which a filler metal is completely liquid.
- Melting range: The temperature range over which the filler metal melts.
- Eutectic: An alloy (filler metal) with a single melting point rather than a melting range is known as a eutectic alloy.

Steps to Successful Brazing

Step 1: Ensure Good Fit and Proper Clearances

Brazing uses capillary action to distribute molten filler metal between the surfaces of the base metals. So when you're brazing, maintain a clearance between the base metals to allow capillary action to work most effectively. In almost all cases, this means a close clearance. Optimal clearance or joint gap for most filler metals is 0.0015 inch, but typical clearances range from 0.001 to 0.005 in.

In everyday brazing, clearances don't have to be overly precise to obtain a sufficiently strong joint. Capillary action operates over a range of clearances, so you have a certain amount of leeway. In everyday shop practice, an easy slip fit typically will produce an adequate brazed joint between two tubular parts. Keep in mind that generally, as the clearance increases, joint strength decreases. Capillary action stops around 0.012 in. If you're joining two flat parts, you can rest one on top of the other. The metal-to-metal contact is all the clearance you'll usually need, because the average mill finish of metals provides enough surface roughness to create capillary paths for the flow of molten filler metal. Highly polished surfaces, on the other hand, tend to restrict filler metal flow.

When you're planning your joint clearances, remember that brazed joints are made at brazing temperatures, not at room temperature. Take into account the coefficient of thermal expansion of the metals being joined, particularly with tubular assemblies in which dissimilar metals are joined.

How much allowance you should make for expansion and contraction depends on the nature and sizes of the metals being joined and the configuration of the joint. Although many variables are involved in pinpointing exact clearance tolerances for each situation, keep in mind this principle: Different metals expand at different rates when heated.

Step 2: Clean the Metals

Capillary action works properly only with clean metal surfaces. If they're coated with oil, grease, rust, scale, or dirt, you must remove these contaminants or they'll form a barrier between the base metal surfaces and the brazing materials.

Cleaning metal parts seldom is complicated, but you have to do it in the right sequence. Oil and grease should be removed first, because an acid pickle solution aimed to remove rust and scale won't work on a greasy surface. Start by getting rid of oil and grease. In most cases you can do this either by dipping the parts into a suitable degreasing solvent, by vapor degreasing, or by alkaline or aqueous cleaning. If the metal surfaces are coated with oxide or scale, you can remove those contaminants chemically or mechanically. For chemical removal, use an acid pickle treatment. Make sure that the chemicals are compatible with the base metals being cleaned and that no acid traces remain in crevices or blind holes. Mechanical removal calls for abrasive cleaning.

Particularly in repair brazing, where parts may be very dirty or heavily rusted, you can speed the cleaning process by using an emery cloth, grinding wheel, or file or grit blast, followed by a rinsing operation. Once the parts are thoroughly clean, flux and braze them as soon as possible to reduce the chance for recontamination of surfaces by factory dust or body oils deposited through handling.

Certain cleaning mediums will leave a residue and embed in the surface, making it unwettable.

Step 3: Flux the Parts

Flux is a chemical compound applied to the joint surfaces before brazing. Its use, with a few exceptions, is essential in the atmospheric brazing process. This is because heating a metal surface accelerates oxide formation, the result of a chemical reaction between the hot metal and oxygen in the air. If you don't prevent these oxides from forming, they'll inhibit the brazing filler metal from wetting and bonding to the surfaces.

A coating of flux on the joint area shields the surfaces from the air, preventing oxide formation. It also dissolves and absorbs any oxides that form during heating or that was not removed completely in the cleaning process.

You can apply flux in any way to the joint as long as you cover the joint surfaces completely. Flux conventionally is made in a paste, so it's usually most convenient to brush it on. But as production quantities increase, it may be more efficient to apply the flux by dipping: dispensing a premeasured deposit of high-viscosity flux from an applicator gun.

Typically, you apply flux just before brazing if possible so it has the least amount of time to dry out and flake off or get knocked off the parts in handling. Choose a flux that's formulated for the specific metals, temperatures, and conditions of your brazing application.

Step 4: Assemble for Brazing

Once your parts are cleaned and fluxed, hold them in position for brazing. Be sure they remain in correct alignment during the heating and cooling cycles so that capillary action can do its job. If the shape and weight of the parts permit, the simplest way to hold them together is by gravity.

You also can aid gravity by adding extra weight as long as you don't add too much. If you add too much weight, your clearances may not be maintained, and the brazing filler metal may be forced out of the joint area. Also remember that adding weight to the parts increases their mass, and that will increase the time required to raise the parts to the brazing temperature.

Step 5: Braze the Assembly

The actual brazing involves heating the assembly to brazing temperature and flowing the filler metal through the joint. Be sure when you're heating an assembly to brazing temperature that you don't heat it to the base materials' melting point.

First, the heating process: In brazing, apply heat broadly to the base metals. If you're brazing a small assembly, you may heat the entire assembly to the flow point of the brazing filler metal. If you're brazing a large assembly, heat a broad area around the joint. A hand-held torch most commonly is used to braze a single assembly. A variety of fuels—natural gas, acetylene, propane, propylene—can be combusted with either oxygen or air. Keep in mind that both metals in the assembly should be heated as uniformly as possible so they reach brazing temperature at the same time. Keep the torch moving at all times, and do not heat the braze area directly.

To prevent uneven heating, watch the flux. If its appearance changes uniformly, the parts are being heated evenly.

Once you've heated the assembly to brazing temperature, you're ready to deposit the filler metal. In manual brazing, carefully hold the rod or wire against the joint area. The heated assembly will melt off a portion of the filler metal, which will be drawn instantly by capillary action throughout the entire joint area. You may want to add some flux to the end of the filler metal rod—about 2 to 3 inches—to improve the flow. You can add flux either by brushing it on or dipping the rod in flux. On larger parts that require longer heating time, or if the flux has become saturated with oxide, adding fresh flux on the filler metal will help improve the flow and penetration of the filler metal into the joint area.

Molten brazing filler metal tends to flow toward areas of higher temperature. In the heated assembly, the outer base metal surfaces may be slightly hotter than the interior joint surfaces. Take care to deposit the filler metal immediately adjacent the joint. If you deposit it away from the joint, it will tend to plate over the hot surfaces rather than flow into the joint. It's also best to heat the side of the assembly opposite the point where you are feeding the filler metal. The filler metal will have a tendency to follow where the heat is most intense.

Step 6: Clean the Brazed Joint

After you braze the assembly, clean it. Because most brazing fluxes are corrosive, cleaning is essential. Cleaning usually is a two-step operation:

- Remove the flux residues.
- Remove any oxide scale formed during the brazing process by pickling.

Because most brazing fluxes are water-soluble, you can remove residues by quenching the assembly in hot water (120 °F or hotter). Immerse the assembly while it's still hot, but make sure that the filler metal has solidified completely before quenching. The glasslike flux residues usually will crack and flake off. If they're a little stubborn, brush them lightly with a wire brush while the assembly is still in the hot water.

You may have trouble removing flux if you haven't used enough to begin with, or if you overheated the parts during brazing. Then the flux becomes totally saturated with oxides, usually turning green or black. In this case, the flux has to be removed by a mild acid solution.

After you've gotten rid of the flux, use a pickling solution to remove any oxides that remain on areas that were unprotected by flux during the brazing process. Generally, the best pickle to use will be recommended by the manufacturer of the brazing materials you're using.

Brazing Techniques

Brazing Flux

A molten brazing filler metal will only wet and flow over a parent metal if both are substantially free of surface oxide. Simply removing surface oxide before brazing is not effective, since a new oxide layer is rapidly formed on heating. To achieve an oxide free surface, it is necessary either to:

- Remove oxide as it is formed by the use of a suitable brazing flux.
- Prevent oxidation during brazing by heating in a protective atmosphere or vacuum.
- Use a self-fluxing brazing alloy - only applicable when brazing copper to copper.

Brazing fluxes are only designed to remove oxide films. Where other contaminants such as oil, paint and lacquer are present these should be removed before brazing, using either mechanical or chemical methods.

Using a Brazing Flux

It is recommended that flux should be applied as a paste to the parts to be joined.

Powders can be made into pastes by stirring in water until the mixture has the consistency of thick cream. A few drops of liquid detergent added to the mixture will often improve the wetting of flux onto clean parent metals. The flux should be applied to both joint surfaces before assembly. Application of flux after assembly places great demands on the fluidity of the molten flux and its ability to penetrate capillary joints. Flux may be applied in the following ways:

- Brushing is an effective method of applying a thin film of paste to the joint itself and to surrounding component surfaces.
- Brazing flux may be applied by dipping one or more components of an assembly into a container of flux. This is most effective with a paste of a thin consistency.
- Automatic application of flux is possible and can be used for mass production.

- Hot rodding is a technique in which a warm brazing rod is dipped into flux powder and the flux adhering to the rod is transferred to the joint area. This is an effective fluxing method but difficult to achieve good penetration of capillary joints. A refinement on this technique is the use of flux coated brazing rods.

Factors which Affect How a Flux Works

The ability of any flux to perform satisfactorily during a brazing operation will depend on the brazing temperature and time, the parent metals, and the volume of flux applied.

Temperature

To be effective the flux must be both molten and active before the brazing alloy melts, and it must remain active until the brazing alloy flows through the joint and solidifies on cooling. It is good practice to select a flux which is active at least 50 °C below the solidus of the brazing alloy and which is still active at a temperature at least 50 °C above the liquids of the brazing alloy. This will ensure that the flux is effective during the brazing operation.

Time

The flux should remove the oxides on the component and must continue to remove fresh oxide until the completion of the brazing cycle. There is a limit to the amount of oxide that the flux can dissolve; although the longer the heating cycle the more likely it is that the flux will become exhausted. With long heating cycles the use of flux with a higher working range would be recommended. For short rapid heating cycles, it may be possible to use a flux above its recommended maximum working temperature.

Parent Metals

Our fluxes are suitable for use on copper, brass, mild steel and most other common engineering materials. Special purpose fluxes exist for aluminium bronze, stainless steel, tungsten, molybdenum and tungsten carbide. Silver brazing fluxes are not effective on aluminium, magnesium, titanium or their alloys.

Using a Copper-Phosphorus Brazing Rod

When joining copper to copper no flux is required. Heat the inserted component first then heat evenly. Transfer the flame to heat all around the base of the socket. When the components are a dull cherry red colour touch the rod onto the joint. Use the heat from the torch to pull the filler metal into and around the joint. Cool in air after brazing.

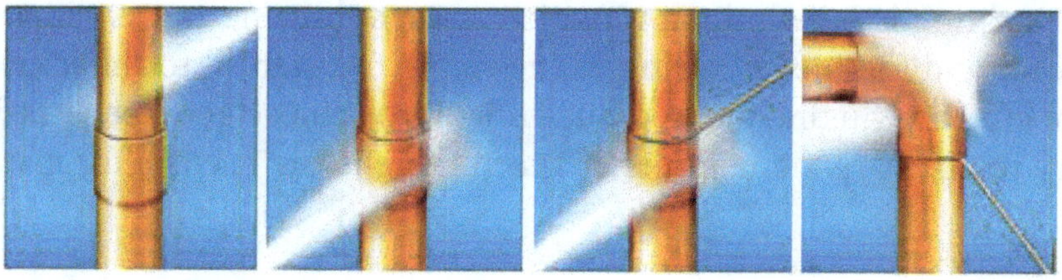

Using a Flux Coated Rod

Flux coated rods combine flux and brazing filler metal in a single convenient form. They are suitable for use on joints where there is a limited gap for the filler metal. As the joint is heated the flux coated rod should be touched or wiped onto the joint area. This will apply some flux and protect the component from oxidation as the joint approaches brazing temperature. At brazing temperature, the now bare end of the rod should again be touched onto the joint and held in position if the alloy begins to flow. Turning the rod between the fingers will ensure that the flux melts off the rod and onto the component evenly. Avoid heating the rod directly with the torch flame as this will melt the flux off the rod leaving no flux available for the next joint. Heat evenly to brazing temperature. Feed the bare end of the flux coated rod into the joint whilst heating the component evenly. Flux residues must be completely removed by soaking in warm water for about 30 minutes.

Using Brazing and Soldering Paste

The use of paste requires different techniques and general practices to those employed for other forms of brazing or soldering consumables. The basic principles always apply.

Location of Paste Deposits

Whilst no particular joint design is excluded from brazing or soldering with paste, the suitability of a component is dependent on there being a place to deposit the required amount of paste. Ideally the component should have a shoulder or ledge on which the paste can be adequately supported.

Components should be assembled and the paste applied as near to the mouth of the joint as possible. Avoid placing paste inside the joint as it may not burn off completely, restricting the flow of the flux and alloy.

If the component does not have a convenient shoulder or ledge then the paste should be applied above the joint. Any movement of the paste during heating can be controlled to direct the flux and alloy flow towards the mouth of the joint. In this case a flux binder system with minimal slump characteristics is recommended.

Dispensing of Paste onto Components

A major advantage of using a paste is that it can be dispensed in accurate and controlled volumes, ensuring that a particular joint receives the same amount of paste on each application.

Paste is often supplied in cartridges, which can be fitted to accurate air operated dispensing machines. Where larger volumes of paste are required a reservoir connected to a paste gun can be used. Paste can be supplied in hand cartridges for manual dispensing although this reduces the close control over paste volumes achieved with even the most basic dispenser.

Contact us about the most suitable method for dispensing paste onto a particular component; we can then formulate the paste accordingly.

Size and Shape of Paste Deposits

The shape and design of the component govern the size and placement of the paste deposit. It is often possible to calculate an approximate amount of paste required per joint. However, in most cases a simple production test will quickly establish the correct amount to use.

Automatic Dispensing

If possible, paste should be applied as a single deposit. This method needs the minimum of dispensing equipment and the least time spent on the dispensing operation.

On long joints, it may be necessary to apply more than one paste deposit. This increases the area covered by flux and alloy and is advisable particularly if the alloy is not free flowing.

Paste can be applied as a number of dots, or as a continuous stripe. In applications such as this, movement of either the dispensing nozzle or the component itself may be required during paste application.

Heating Methods for Brazing and Soldering Paste

Our paste products can be used successfully with most of the standard heating methods:

- Fixed burners: Accurate and reproducible heat patterns can be set electronically using a range of fuels. They may be arranged as single or double burners on basic shuttle or rotary indexing machines.

- Furnace heating: Uses a reducing atmosphere or vacuum to prevent or break up surface oxides during the brazing process.
- Induction heating: Quick and easily controlled heating well suited to paste products.
- Resistance heating: Works well with brazing and soldering pastes. Resistance heating with a separate filler metal and flux can cause problems because the flux acts as an insulator. Using paste will help to overcome this as the finely divided powder offers a conductive path through the flux.
- Hand torch: Pastes should be heated indirectly to ensure the paste deposit reaches the correct temperature at the same rate as the rest of the component. If heated directly, the flux in the paste will not have sufficient time to clean the parent metals and the molten filler metal alloy will not wet out onto the joint area.

Heating Brazing Paste

The behaviour and flow characteristics of a brazing paste when heated will depend upon its formulation, whether it contains flux and the heating method being used. A typical fluxed paste for brazing in air with a silver brazing filler metal based around the 'B1' binder system will go through the following stages:

- On initial heating, the paste deposit will increase in size. During this early stage of heating the paste deposit should be heated slowly and indirectly to cause it to 'set'. Heating it too rapidly or directly can cause it to 'explode' or spit.
- With further heating, the paste will either start to smoke, or if a flame is present ignite. Where ignition of the paste does not occur, for example when using RF induction or resistance heating, local exhaust ventilation should be used to remove the fume from the working environment.
- As the heating continues the paste's appearance will change from a shiny, glossy deposit, to a dull, dry one. The binder has now been lost from the paste which has

become 'set'. Once 'set' it is stable and can be subjected to a more rapid rate of heating.

- Next the flux starts to melt, 'wetting' the parent metals and reducing surface oxides on them. Initially this takes place locally around the base of the paste deposit. As the temperature continues to increase the flux becomes more fluid, spreading out and flowing into the capillary gap present within the joint.
- Finally, as more heat is applied the filler metal will begin to melt and then flow completely. The rate of heating at this stage needs to be sufficient to prevent liquidation of the filler metal. This is where the filler metal only partially melts leaving behind a skull of solid material and resulting in a joint that is only partially filled.

Both the binder and the flux in a paste can be changed or modified to match the needs of a particular brazing job. For example, binder systems can be supplied that collapse and spread onto a component where paste is needed over a large area.

Removal of Brazing Paste Residues after Heating

The flux residues of brazing pastes are corrosive and therefore their removal after brazing is essential. They are similar to those generated in other brazing operations and may be removed by the same methods - soaking in hot water (> 40 °C for 30 minutes), soaking in 10% sulphuric acid or by mechanical removal (e.g. grit blasting). The method should depend on the type of flux present. Brazing pastes often leave a 'footprint' or mark on the component that is difficult to remove after brazing.

Heating Soldering Paste

Solder pastes are formulated around totally different binder systems and flux types to those used for brazing and they therefore react completely differently when heated.

Solder paste when heated initially will exhibit some slump depending upon its formulation. With further heating the deposit may bubble and heave (foam) and begin to smoke. Local exhaust ventilation should be used to remove the fumes. If heated too rapidly the paste deposit can boil and spit. Gentle, indirect heating (such as hot air heating) is recommended throughout the soldering process.

The solder paste becomes more fluid as the temperature is increased but it should remain colourless throughout the soldering operation. If the flux starts to go brown it is an indication that either it is beginning to become exhausted or that the paste has been overheated.

As the solder alloy within the paste begins to melt, the flux is displaced from the joint and floats on top of the molten solder. Fluxes and solders will always flow to the hottest point of a joint and further flow of the solder into, along or around the joint can be encouraged by creating a thermal gradient across it.

If the solder fails to wet and flow as required or forms a molten 'ball' then this is an indication that the flux has not been able to remove the oxides present on the surface of the parent materials. This could be due to exposing the paste deposit to too high a temperature or exhausting the flux before the filler metal becomes molten. Alternatively, an inappropriate fluxed binder system was selected for the application.

Removal of Soldering Paste Residues after Heating

Flux residues from soldering paste are classified as non-corrosive, intermediate or corrosive. Non-corrosive residues may be left on the completed joint. Intermediate or corrosive flux residues should be fully removed. Warm water, mildly alkaline solution, or in the case of rosin based flux residues an organic solvent, should be used.

Storing and Shelf Life

Brazing and soft soldering pastes have limited shelf life and strict stock rotation should be exercised during their storage. The binder system and the filler metal powder can separate and the paste may dry out during storage. Separation or drying out of a paste is more likely if the product is not stored correctly.

Storage Temperature

Braze and soft solder pastes should not be stored below 5 °C or above 25 °C. The ideal storage temperature for pastes is between 10 and 15 °C. The viscosity of brazing and soft solder pastes is affected by temperature. They are designed for use between 18 and 24 °C (Viscosity measurements are taken during production at 20 °C). If subjected to temperatures below 15 °C the paste viscosity will increase making dispensing more difficult. If stored at temperatures above 25 °C the paste will become very fluid and may separate and dry out. It is therefore recommended that pastes should be withdrawn from the storage facility some time before use and placed in the workshop environment to obtain a normal working temperature. Storing at temperatures below 5 °C can result in damaging the products irreparably.

Storage Conditions

- Paste should ideally be stored in cool dry conditions away from direct sunlight and other sources of heat.
- Pastes supplied in buckets or pots should be stored in closed containers and stood with the lid upper most.
- Pastes supplied in machine cartridges or hand syringes should be stored standing up in racks with the 'follower plug' upper most and the nozzle facing down. Storing cartridges/syringes on their side can result in separation of the paste.

Soldering

Soldering is the process of joining two or more electronic parts together by melting solder around the connection. Solder is a metal alloy and when it cools it creates a strong electrical bond between the parts.

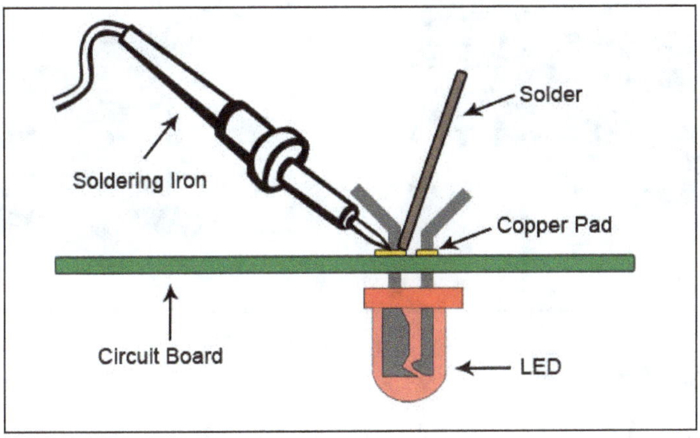

Soldering Tools

Soldering Iron

A soldering iron is a hand tool that plugs into a standard 120v AC outlet and heats up in order to melt solder around electrical connections. This is one of the most important tools used in soldering and it can come in a few variations such as pen or gun form. For beginners, it's recommended that you use the pen style soldering iron in the 15W to 30W range. Most soldering irons have interchangeable tips that can be used for different soldering applications. Be very cautious when using any type of soldering iron because it can heat up to 896′ F which is extremely hot.

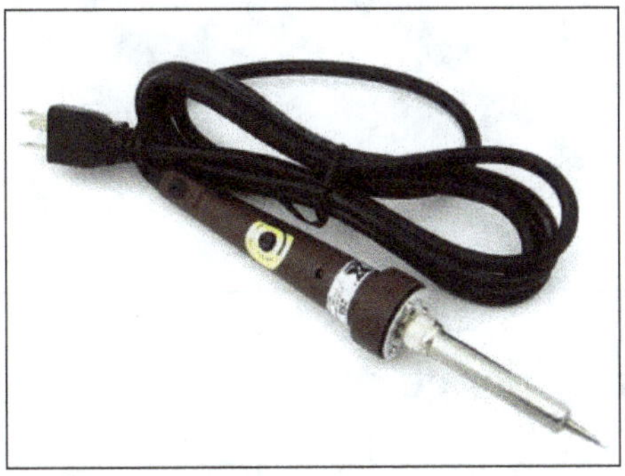

Soldering Station

A soldering station is a more advanced version of the basic standalone soldering pen. If you are going to be doing a lot of soldering, these are great to have as they offer more flexibility and control. The main benefit of a soldering station is the ability to precisely adjust the temperature of the soldering iron which is great for a range of projects. These stations can also create a safer workspace as some include advanced temperature sensors, alert settings and even password protection for safety.

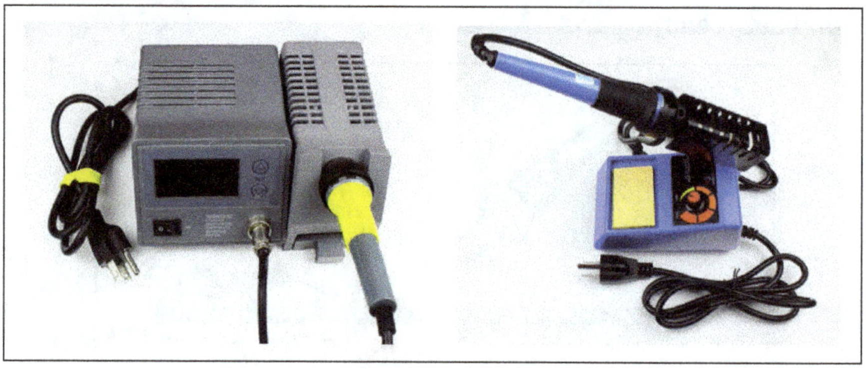

Soldering Iron Tips

At the end of most soldering irons is an interchangeable part known as a soldering tip. There are many variations of this tip and they come in a wide variety of shapes and sizes. Each tip is used for a specific purpose and offers a distinct advantage over another. The most common tips you will use in electronics projects are the conical tip and the chisel tip.

- Conical Tip: Used in precision electronics soldering because of the fine tip. Because of its pointed end, it's able to deliver heat to smaller areas without affecting its surroundings.
- Chisel Tip: This tip is well-suited to soldering wires or other larger components because of its broad flat tip.

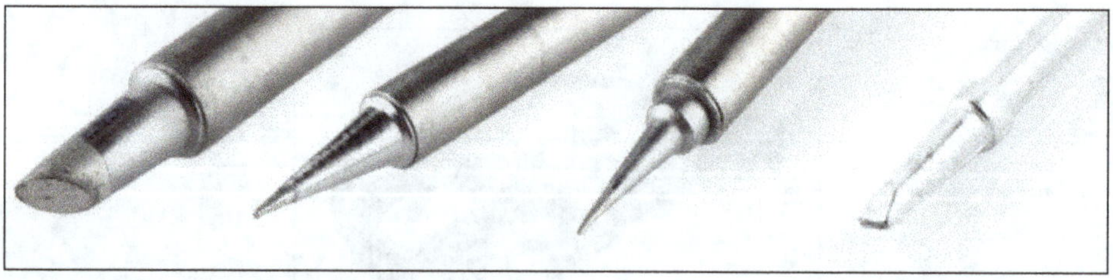

Brass or Conventional Sponge

Using a sponge will help to keep the soldering iron tip clean by removing the oxidation that forms. Tips with oxidation will tend to turn black and not accept solder as it did when it was new. You could use a conventional wet sponge but this tends to shorten the lifespan of the tip due to expansion and contraction. Also, a wet sponge will drop the temperature of the tip temporarily when wiped.

Soldering Iron Stand

A soldering iron stand is very basic but very useful and handy to have. This stand helps prevent the hot iron tip from coming in contact with flammable materials or causing accidental injury to your hand. Most soldering stations come with this built in and also include a sponge or brass sponge for cleaning the tip.

Solder

Solder is a metal alloy material that is melted to create a permanent bond between electrical parts. It comes in both lead and lead-free variations with diameters of .032″ and .062″ being the most common. Inside the solder core is a material known as flux which helps improve electrical contact and its mechanical strength.

For electronics soldering, the most commonly used type is lead-free rosin core solder. This type of solder is usually made up of a Tin/Copper alloy. You can also use leaded 60/40 (60% tin, 40% lead) rosin core solder but it's becoming less popular due to health concerns. If you do use lead solder, make sure you have proper ventilation and that you wash your hands after use.

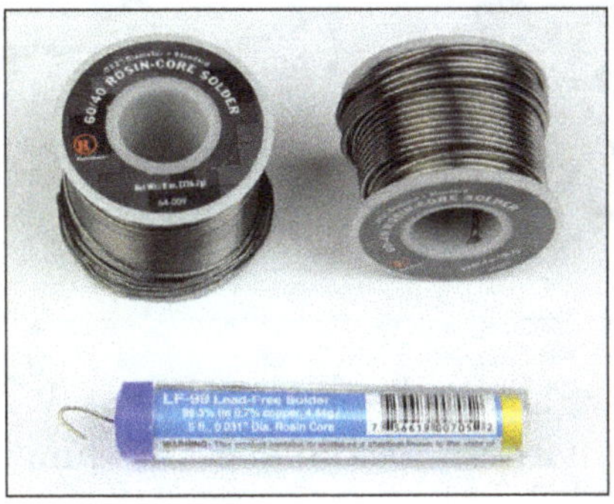

When buying solder, make sure not to use acid core solder as this will damage your circuits and components. Acid core solder is sold at home improvement stores and is mainly used for plumbing and metal working.

Solder does come in a few different diameters. The thicker diameter solder (.062″) is good for soldering larger joints more quickly but it can make soldering smaller joints difficult. For this reason, it's always a good idea to have both sizes on hand for your different projects.

Helping Hand

A helping hand is a device that has 2 or more alligator clips and sometimes a magnifying glass/light attached. These clips will assist you by holding the items you are trying to solder while you use the soldering iron and solder. A very helpful tool to have in your maker space.

Soldering Safety

Soldering irons can reach temperatures of 800′ F so it's very important to know where your iron is at all times. We always recommend you use a soldering iron stand to help prevent accidental burns or damage.

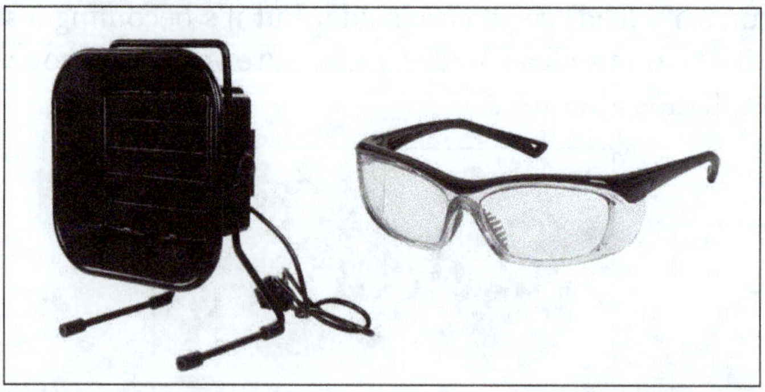

Make sure you are soldering in a well-ventilated area. When solder is heated, there are fumes released that are harmful to your eyes and lungs. It's recommended to use a fume extractor which is a fan with a charcoal filter that absorbs the harmful solder smoke you can visit sites like Integrated Air Systems for air filtration systems.

Steps for Soldering

Wetting

When the hot solder comes in contact with a copper surface, a metal solvent action takes place. The solder dissolves and penetrates the copper surface. The molecules of solder and copper blend to form a new alloy, one that's part copper and part solder. This solvent action is called wetting and forms the intermetallic bond between the parts.

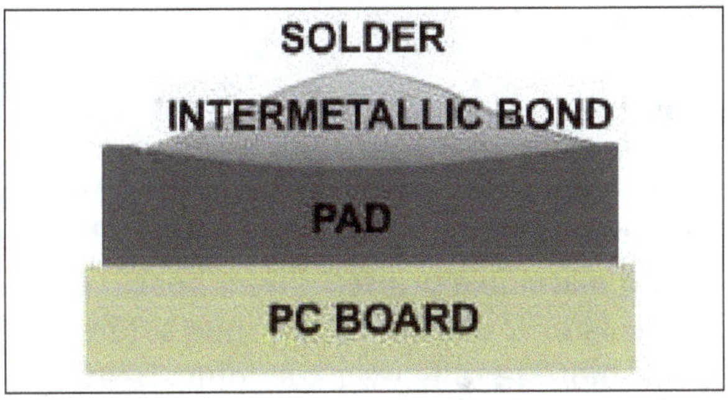

Wetting can only occur if the surface of the copper is free of contamination and from the oxide film that forms when the metal is exposed to air. Also, the solder and work surface need to have reached the proper temperature.

Tinning

New tips must be tinned immediately when used for the first time. Tinning components involves applying a thin coat of solder to the surface to be soldered. This greatly helps in achieving a good joint and should be done whenever possible. A continuously tinned surface must be maintained on the soldering iron tip's working surface to ensure proper heat transfer and to avoid transfer of impurities to the solder connection.

It's better to re-apply a very small amount of solder again, mainly to improve the thermal contact between the iron and the joint, so that the solder will flow more quickly and easily. Insert the components and splay the leads so that the part is held in place.

De-Oxidation

Scrape the Printed Circuit Board and lead of the component using steel wool to remove oxidation. Use a small hand-held file or scrape using a knife blade or rub a fine emery cloth over them to reveal fresh metal underneath. After preparing the surfaces, avoid touching the parts afterwards if at all possible. To form the component properly checks whether the component easily goes inside the PCB.

Flux

Flux removes oxides which arise during heating, and is seen as a brown fluid bubbling away on the joint. Flux is a solvent consisting of natural or synthetic rosins to remove oxides on the copper surface. It is non-corrosive and non-conductive and will not affect the circuitry. To make any good solid joints, the components have to be both physically and chemically clean as the solder needs to actually get into the components not just on them. Scraping the leads of the components and dipping the leads in flux and apply flux on the pads of the PCB also helps to remove the surface oxides in the soldering process.

Temperature Control

Another step to successful soldering is to ensure that the temperature of all the parts is raised to roughly the same level before applying solder. Imagine, for instance, trying to solder a resistor into place on a printed circuit board. it's far better to heat both the copper PCB and the resistor lead at the same time before applying solder, so that the solder will flow much more readily over the joint. Heating one part but not the other gives a far less satisfactory joint, so strive to ensure that the iron is in contact with all the components first, before touching the solder to it. The melting point of most solder is in the region of 188 °C (370 °F) and the iron tip temperature is typically 330-350 °C (626°-662 °F). The latest lead-free solders typically require a higher temperature.

Time

The joint should be heated with the bit for just the right amount of time - during which a short length of solder is applied to the joint. Do not use the iron to carry molten solder over to the joint. Excessive time will damage the component and perhaps the circuit board copper foil too. Heat the joint with the tip of the iron, then continue heating whilst applying solder, then remove the iron and allow the joint to cool. This should take only a few seconds, with experience. The heating period depends on the temperature of your iron and size of the joint - and larger parts need more heat than smaller.

Applying Solder

In general, the soldering iron tip should be applied to the maximum mass point of the joint. This will permit the rapid thermal elevation of the parts to be soldered. Molten solder always flows from the cooler area toward the hotter one. Before solder is applied, the surface temperature of the parts being soldered must be elevated above the solder melting point. Never melt the solder against the iron tip and allow it to flow onto a surface cooler than the solder melting temperature.

To solder, place the tip of the soldering gun directly against the area to be soldered and carefully apply the hot solder to the area where the soldering gun touches the metal. Only apply the minimum amount of solder needed to set the connection. Never blow on the hot solder or try to use the connection before it is completely cooled. This will create small fractures in the metal joint and weaken the bond between pieces.

Solder Coverage

The final key to a successful solder joint is to apply an appropriate amount of solder. Too much solder is an unnecessary waste and may cause short circuits with adjacent joints. Too little and it may not support the component properly, or may not fully form a working joint. How much to apply, only really comes with practice. A few millimeters only, are enough for an "average" PCB joint.

Post Solder Cleaning

When cleaning is required, flux residue should be removed as soon as possible, but no later than one hour after soldering. Some fluxes may require more immediate action to facilitate adequate removal. Mechanical means such as agitation, spraying, brushing, and other methods of applications may be used in conjunction with the cleaning solution.

Re-soldering

Care should be taken to avoid the need for re-soldering. When re-soldering is required, quality standards for the re-soldered connection should be the same as for the original connection. A cold or disturbed solder joint will usually require only reheating and reflowing of the solder with the addition of suitable flux. If reheating does not correct the condition, the solder should be removed and the joint re-soldered.

Effective Soldering

Making the Perfect Solder Joint

- All parts must be clean and free from dirt and grease.
- Try to secure the work firmly.
- "Tin" the iron tip with a small amount of solder. Do this immediately, with new tips being used for the first time.
- Clean the tip of the hot soldering iron on a damp sponge.
- Many people then add a tiny amount of fresh solder to the cleansed tip.

- Heat all parts of the joint with the iron for under a second or so.
- Continue heating and then apply sufficient solder, to form an adequate joint.
- Remove and return the iron safely to its stand.
- It only takes two or three seconds at most, to solder the average PCB joint.
- Do not move parts until the solder has cooled.

Characteristics of a Good Connection

A good solder joint has very few things to look for compared to a poor one. A good solder joint shows the following characteristics:

- Smooth
- Bright
- Shiny
- Clean
- Concave solder fillet
- Good wetting
- The end of the wire or lead is covered with solder

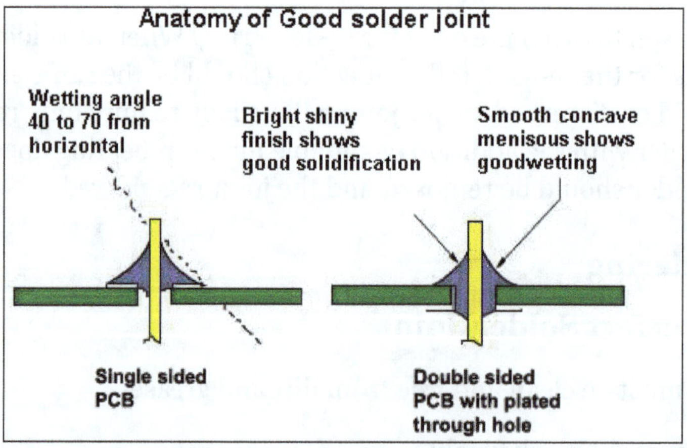

Welding Process for Plastics

Welding of plastics is widely used in a number of industries particularly for joining of thermoplastic films and sheets. All welding processes employed at present involve the application of heat to the area of contact. According to the source of heat employed the welding process for plastics may be divided into two broad classes as shown in figure.

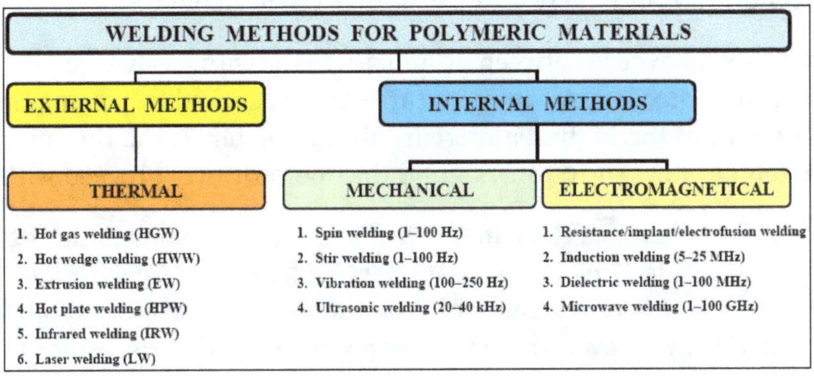

Figure: Welding Process for thermoplastics polymers.

Direct Utilisation of Heat Sources

One class of welding processes utilises heat from an extraneous source such as a stream of hot gas, a hot extruded filler material, or a hot tool. In all these processes heat is transferred to the surfaces being welded by conduction, convection, and radiation.

The second group includes processes in which heat is generated within the workpiece through conversion of some other form of energy such as high frequency current, ultrasonic waves, friction, infra-red light, chemical reactions, or neutron irradiation.

The mechanism of welding of plastics is considered to be the phenomenon of auto-cohesion by which welding is accomplished by the diffusion of some molecular chains from one piece into another to form a strong macro-molecular bond between the two pieces.

Welding of plastics is done in the viscous fluidic state under the application of pressure. Better weld ability is shown by thermoplastics which have a wider softening range rather than a sharp melting point. Because the coefficient of thermal expansion of plastics is several times that of metals, residual stresses may develop in the weldment resulting in reduced joint strength.

The factors affecting process selection for welding of plastics include workpiece thickness, physio-chemical properties of the plastic, design of the article, and the number of components to be produced. The filler material used in welding plastics should be as close in mechanical properties to the parent material as possible.

Hot Gas Welding

In this process a jet of hot gas which may be air, nitrogen, argon, products of combustion of some fuel gas (for example, acetylene, hydrogen, LPG) is played on the edges to be joined as shown in figure. Fuel gas cannot be used directly to weld plastics because the flame has a very high temperature.

That is why special torches have been developed for the hot gas welding of plastics. The welding gas may be heated by electricity or by flame. Air temperature can be adjusted by varying its flow rate and the resistance of the electric element.

The flow rate is set anywhere between 25 and 30 m/sec with a valve, and the resistance of the circuit with a rheostat. Air temperature can be checked by placing the tip of the torch within 5 mm of the bulb of a mercury thermometer. If the thermometer reads the required temperature in 10 to 15 second the operator may proceed with welding.

For operator safety, electrically heated torches operate on voltage not exceeding 36 volt. The efficiency of electrically heated torches is 60 percent. Such torches are simple to make and there is no open flame therefore these may be used in room holding inflammable materials. However, these torches are heavy and, thus, rather unwieldy for use in places difficult of access or in awkward positions.

Gas torches may be either directly or indirectly heated. In directly heated gas torches the welding gas is mixed with the products of combustion of fuel gas while in indirectly heated gas torches the products of combustion transfer their heat to the welding gas through the wall. The fuel gas (C_2H_2, H_2 LPG, etc.) is used under a pressure of 0.5 to 10 N/cm².

In comparison with electrically heated torches gas torches can weld at a higher rate, are lighter in weight and more durable. When used eight hours a day, the service life of a gas torch is 1.5 to 2 years. A major drawback of gas torches is that the gases used are inflammable and explosive.

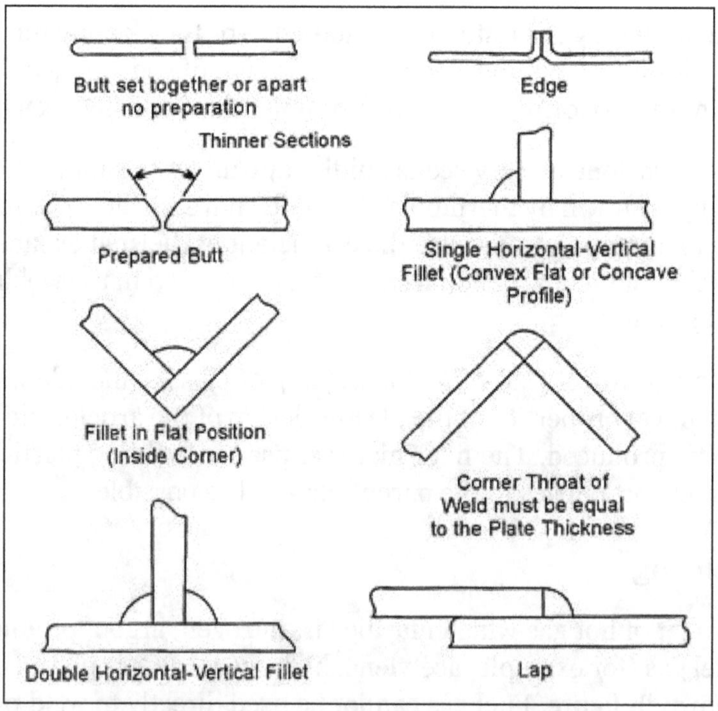

Figure: Edge preparation for butt joints in plastics.

Joint Design

Usually butt joints are preferred because lap, tee, and fillet joints are more difficult to make. Depending upon the work thickness, square edge, single vee, and double vee edge preparations are employed for butt joint preparation as shown in figure. The standard edge preparation for butt welds requires a root gap but no root face.

Double vee joints are usually stronger than single vee joints and the groove angle has a decisive effect on joint strength. As a rule joint strength increases as the groove angle is increased because better penetration is obtained at the root; however the production rate is lowered.

Welding Procedure

The fusion faces are carefully cleaned and de- greased,-say with acetone; the glossy spots are removed with emery paper or scraper. Before welding torch is switched on or fired the welding gas is turned on and its flow rate adjusted. The gas is then fired in case of a gas torch or electric current switched on for an electric torch.

Filler rods used come in diameters of 2, 3, 4, ± 0.5 mm and other shapes like triangular and trapzoidal of different sizes. The filler rods are fabricated from the same material as the work material but may be have different colour and usually contain higher percentage of plasticizer to lower down its softening point.

The filler rods may either be cut to lengths of at least 0.5m or tied up in bundles or uncut and supplied in coils of 3 to 4 kg. The size of filler rod is chosen to suit the work thickness, type of edge preparation and the strength desired. Thicker rods usually result in reduced joint strength.

The torch tip size is selected depending upon work thickness, and edge preparation. Tips with orifice diameter of 1-5 – 2 mm are used to weld sheets 3 to 5 mm thick while tips with diameter of 3-5 – 4 mm are used for sheets 16 to 20 mm thick. As a rule the orifice diameter of a tip should be equal to the diameter of filler rod used. Otherwise the rod will not be heated adequately and the strength of the joint will be impaired.

The torch-to-work angle is chosen in relation to material thickness. For sheet under 5 mm this angle should preferably be 20°-25° and for sheets in the range of 10—20 mm, it should be 30°- 45°. The torch tip-to-work distance should be kept constant between 5-8 mm. To produce a good bond between the filler and the work the rod should be heated and fused at the start of the weld so that its end extends 3—5 mm beyond the end of the work.

The hot gas stream must not be directed in any one position, instead it should be moved continuously over a short length of the welding rod and the surface being welded so as to heat both uniformly. The joint edges and the filler rod are rapidly heated at the surface because plastics are poor conductors of heat.

However, it is essential to heat the rod throughout its bulk so that it can be thoroughly softened at the centre and properly placed in the groove. That is why thicker rods cannot be used and welding is slow by hot gas process particularly with filler rod technique. If the pressure is not applied properly the softened rod is compressed in the direction opposite to its motion which produces waviness in it.

The filler rod should be fed square to the weld so as to exert proper control of pressure. As the filler rod is forced down by hand it welds to the softened edges and forms a weld, as shown in figure.

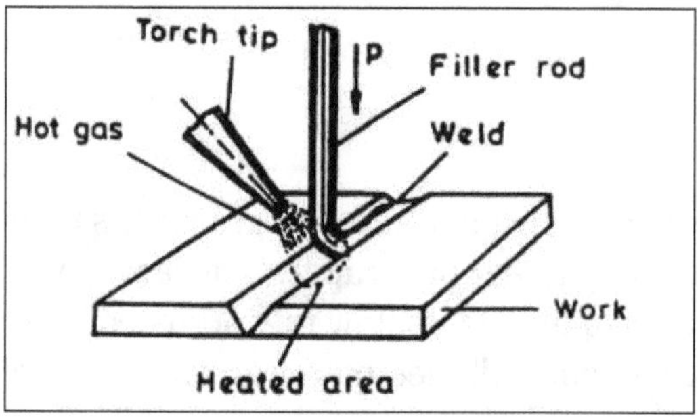

Figure: Manual hot gas welding using filler material.

Hot gas welding without filler rod accelerates the process and enhances the mechanical properties of the joint. A simple setup for this technique is shown is figure. In this method the sheet edges are scarfed and fitted before being uniformly heated by hot gas.

The hot gas jet is followed by cold rollers which exert the required pressure to complete the weld. Welding rate with this technique may be 12 to 20 m per hour, depending upon sheet thickness. The strength of the joint is 80 to 90 percent that of the parent material and the impact strength remains the same. Hot gas welding without filler material is most often applied to make lap joints in films. For critical joints it is better to seal the weld root to improve the joint strength and quality.

After welding the joint is allowed to cool. Artificial cooling particularly in material thicker than 10 mm may lead to cracking.

The strength of butt welds in plastics is 65% that of the parent material in shear, 75% in tension, 85% in compression, and 65% in bending while that of fillet welds is 65% in tension. The impact strength of the weld material is usually very low.

Apart from low strength of the joint hot gas welding also results in reduced plasticity in the weld and near-weld area, low production rate especially in thick sheets, danger of overheating and dependence on the operator skill. In spite of these limitations hot gas welding is widely used for welding PVC, polyethylene, acrylics, and polyamide.

For welding PVC, hot gas welding process is most often used. PVC has no sharp melting point. At a temperature of over 80 °C it softens. At 180 °C it begins to flow, and at 200 – 220 °C it passes into viscous fluidic state; if pressure is then applied it will weld. The welding temperature must be kept below the critical point at which the material begins to decompose.

To obtain an optimal temperature of 200 – 220 °C for hot air in the welding zone, it should be heated to 230 – 270 °C in the torch. The effect of air temperature on welding rate and joint strength are presented in table.

Table: Effect of air Temperature and welding rate on joint strength in PVC welds.

Air temperature (°C).	Welding rate for single run (m/h)	U.T.S For double vee joint (N/mm²)	Joint efficiency (%)	Remarks
210	4.8	14	25	Indirectly hated gas torch.
230	8.4	34	25	Air pressure 0.6 Atmosphere
250	11.4	32	58	Tip diameter, 2.5mm
270	13.8	35	63	Rod diameter, 2.5mm
300	15.0	17	30	
320	Material	Decomposes		

If a correct welding temperature has been chosen, a dull spot appears on the PVC sheet 2 or 3 seconds after the jet of hot air has been played on it.

Weld quality in PVC depends upon the rate at which the filler rod is fed into the joint, the angle at which it is fed into the joint, the force applied to press the heated rod into the joint, the distance of the torch tip from the work surface, the position and direction of the torch during welding. A filler rod 3 mm in diameter should be fed to the joint at a rate of 12 to 15 m per hour.

Welds made in PVC by hot gas filler rod technique show low impact strength. PVC is sensitive to stress concentration to such an extent that even when a rod is welded to a tube the impact strength of the joint is just about 10% of the impact strength of the parent material.

The welding of PVC by hot gas welding is a slow process. For example to weld one metre of PVC, 18—20 mm thick, with V edge preparation it is necessary to lay 30 to 35 rods, 3 mm in diameter, requiring about 2 hours to accomplish the job. The welding speed can be increased by raising the gas temperature to 300 °C and by preheating the filler rod but this requires careful monitoring of the process otherwise the higher temperature may lead to the decomposition of the material.

Acrylics are welded with a jet of air of 200 – 220 °C. Time taken for welding acrylic sheet is almost double that required for PVC sheet of the same thickness, and the welding rate is therefore nearly halved. The filler rods used are cut from acrylic sheet and have a cross-sectional area of 7—12 mm². Acrylics can also be welded satisfactorily by using PVC filler rods. To achieve quality welds in acrylics it is best to degrease the surfaces to be welded with acetone or dichloromethane prior to welding. The tensile strength of welded joints in acrylics is generally 3P – 45 % that of the parent material.

Polyethylene should preferably be welded with N_2 or CO_2 gas heated to 200— 220 °C, although gas flame torches may also be used. Hot gas welding is also frequently used to weld vynil plastics, polystyrene, and some other plastic materials.

The major use of hot gas welding is in the production of very large fabrications made from sheet materials, for example, ducting, and pipe work and ventilator hoods for chemical plant installations. This method is normally not used for joining small parts.

Extruded- filler Welding

In this method the filler in a viscous fluidic state is fed into the joint. The hot filler material melts the edges of the plastic being joined and a strong bond is formed between the filler and the parent material. In a way this process resembles the hot gas process with filler rod technique. Satisfactory welds can be accomplished by this process both in films and heavy gauge sheets.

Hot Tool Welding

This process can be performed by several techniques depending upon the type of tool employed which may include hot blade, hot wedge, hot plate, strip heater or a press.

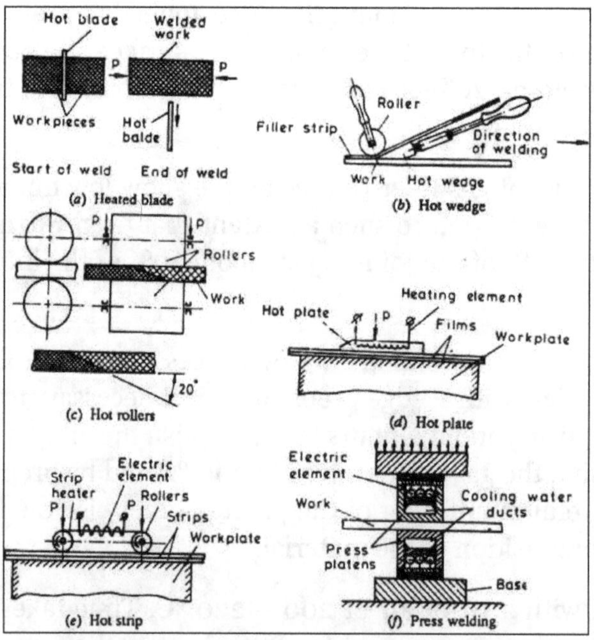

Figure: Typical methods for hot tool welding of plastics.

In hot blade welding, the heated blade is placed between the surfaces to be joined as shown in figure. After the hot blade has softened the surfaces it is rapidly withdrawn and the surfaces are brought in contact under pressure to accomplish the weld. This process can be used to make butt and lap joints over the entire surface of contact at the same time.

In hot wedge welding shown in figure the heated wedge is place between the surfaces to be joined and is moved along the line of welding as the edges are softened. Pressure is applied through a roller to the top strip to weld it to the bottom sheet.

This process is used to weld elastic materials but can also be used to weld thin rigid sheets or straps upto 5 mm to thicker sheets. Precautions are, however, needed in this process to avoid sticking of work material to the hot wedge. Best of all this process can be used for welding films by using pressure rollers arranged above and below the films being joined together as shown in figure. Apart from hot wedge method films can also be welded by hot plate, hot strip and thermal impulse methods.

In hot plate welding, resistance heated plate is moved over the films to be lap welded. When the desired welding temperature is reached, pressure is applied to accomplish the weld. The films to be welded are laid out on a work plate as shown in figure.

In hot strip welding the strip heater, heated by an electric element, is advanced by rollers and is simultaneously forced by pressure P against the films to be lap welded which are laid out on a workplate as shown in figure. The films can be advanced under the pressure rollers by moving either the welding head or the workplate.

In thermal-impulse process the material (films) is raised to the welding temperature almost instantaneously as a strong current pulse is passed through an electric heater. The heater may be of point, strip, or even an odd shaped type. Because the heat can be accurately metered, overheating at the joint is avoided.

In press welding heat is transferred to the area to be welded by the hot platen of the welding press. The plastic pieces with their edges scarfed are clamped between the resistance heated press platens as shown in figure. After the workpieces have been raised to the welding temperature, they are kept under the required pressure as the platens are cooled by the water circulated through the ducts.

Presses usually make butt welds. A typical plastics welding press for butt joints develops fairly high pressures, heats the work locally, and compresses the softened zone from all sides. That is why this technique is also referred to as static-jig welding. This technique can butt weld sheets, bars, strips and plates.

Stresses may be developed in welding of plastic especially if the sheets to be welded are large in thickness. To relieve these stresses it is a good practice to anneal the welded articles from a temperature 25 to 30 °C below the softening point of the material.

Hot tool welding produces strong welds at a high production rate. This process is applicable to plastics which cannot be joined by high frequency induction welding, for example PTFE (polytetrafluorethylene), polyethylene, and polystyrene. Butt, fillet and T-joints can be made by this process. Acrylics joined by hot tool welding retain transparency and clarity at and around the joint it can also be used for welding films for seams of considerable length. When large quantities of welds are required, the hot tool welding method can be easily mechanised.

Conversion of Energy into Heat

High Frequency Induction Welding

In H.F. induction welding the workpiece is placed in a high frequency field set up between two metal electrodes as shown for roller seam welding. Only those plastics which are imperfect dielectric can be welded by this process.

The few free electrons existing in such plastics give rise to conduction current when the material is placed in the H.F. field. The work done to displace the charged particles is converted into heat. Some heat is also generated when the field alternates. To increase the amount of heat generated use is made of very high frequency current in the range of 30 to 40 MHz or even higher. Generally no filler material is used. As all the heat is generated directly in the body of the workpiece being welded, welding speed is high and the electrodes are not overheated.

H.F. induction process is used to make spot, static-jig, and scam welds; however butt, fillet, and tee joints are difficult to make. The welds produced are tight and strong. The process can be easily automated to weld films, sheets, and tubes. Lap welds by seam welding machines can be carried out at speeds as high as 27 to 65 m/hr.

Among the merits of high frequency welding are high production rate, economy, and satisfactory joints. It can weld materials upto 5mm thick. However, materials with low dielectric dissipation factor like PTFE, polyethylene, and polystyrene are not possible to weld by H.F. induction welding. But polyethylene can be welded by this process by placing a strip of PVC in the joint. PVC being an imperfect dielectric gets heated up under the action of H.F. current and transfers the heat to polyethylene to accomplish the weld.

Friction Welding

Plastics are friction welded the same way as metals, though normal setup consists of rotating one piece and keeping the other stationary, as shown in figure, but large pieces can be welded by keeping them stationary and rotating a short insert between them. The quality of the weld depends upon the speed of rotation, the axial force applied, and the amount of plastic deformation involved.

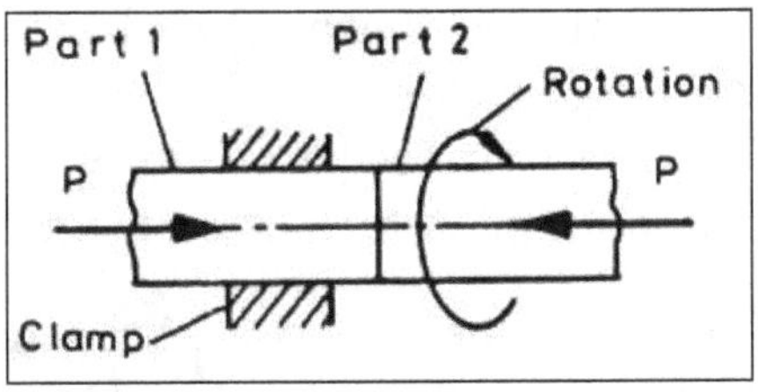

Figure: Friction welding of plastic rods.

Because the heat is generated at the interface the properties of the adjoining material are not affected and the joint has good mechanical properties. Due to heat produced directly on the surfaces being joined this process has the advantage of high welding rate, adaptability to automatic control and usability under field conditions. However, the process can be used only if one of the components is cylindrical so that it can be rotated. Also flash formed at the joint means not only the wastage of material but also the additional cost in machining to remove it.

Friction welding of PVC tubes and pipes is well developed. Prior to welding the ends of the tubes are sized by heating the tube ends in oil to 100 °C for 3 to V4 minutes and then clamping the tubes in gauge for 3 minutes followed by water cooling to room temperature. The welding is accomplished by rotating one of the tubes in a chuck.

The speed of rotation depends on tube diameter, for example, 50 mm diameter tube is rotated at 800 rpm while 80 mm diameter tube is rotated at 600 rpm and the spinning time is 1 ± 0.5 minutes. After the desired viscous fluidic temperature of 140 – 160 °C is reached the rotation is stopped and a pressure of 20 to 40 N/cm² is applied till the weld is cooled to room temperature in about 7 to 10 minutes.

Friction welds in PVC compare in quality with the parent material. Typical joint strength on like materials is about 90 % that of the parent material.

Ultrasonic Welding

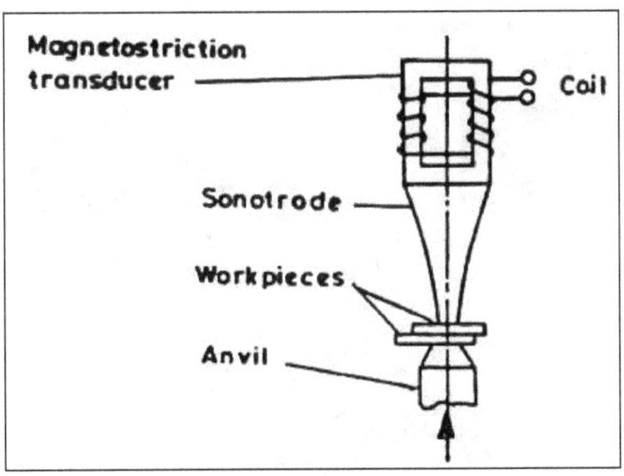

Figure: Ultrasonic spot welding of plastics.

For ultrasonic welding of plastics the welding machine has the same features as the one for metals. The main element of the welding machine is a transducer, which converts the H.F. energy supplied by ultrasonic oscillator into vibrations. The vibrations are applied to the work through a sonotrode which is set up on an anvil as shown in figure.

The mechanical vibrations applied to the work cause the generation of heat in the plastic material. Pressure is applied to the softened material to complete the joint. Welding takes place the same instant as the H.F. voltage is applied to the transducer coil. The frequency used is upto about 20 KHz.

The distinct features of ultrasonic welding include:

- Its capability to veld from one side, that is, no second electrode is required, thus the second piece may be of unlimited thickness.
- The ultra-sonic energy can be applied at a constable distance from the weld being made.
- Due to localisation of heat no overheating of the bulk material takes place.
- Concentration of heat at the desired spot helps in higher welding speed.
- Surface contaminants like grease, graphite, electrolytes have no appreciable effect on weld quality.
- No interference in radio reception.
- No voltage needs to be applied to the welding tool.
- A large number of thermoplastics and a wide range of thickness can be welded by ultrasonic welding.
- Ease of application and automation.

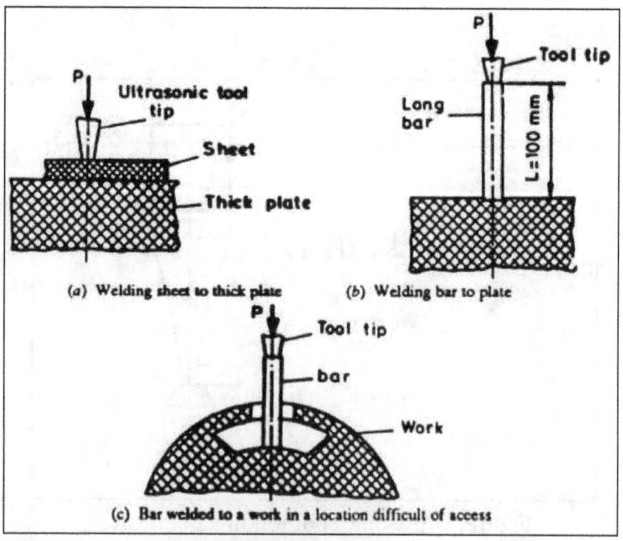

Figure: Specific field of application for ultrasonic welding.

Specific field of application for ultrasonic welding is the spot and jig welding of films and sheets of medium and large thickness, in the sealing of cartons where the seam may be contaminated with grease, and in the packing of various preserves containing current conducting liquids.

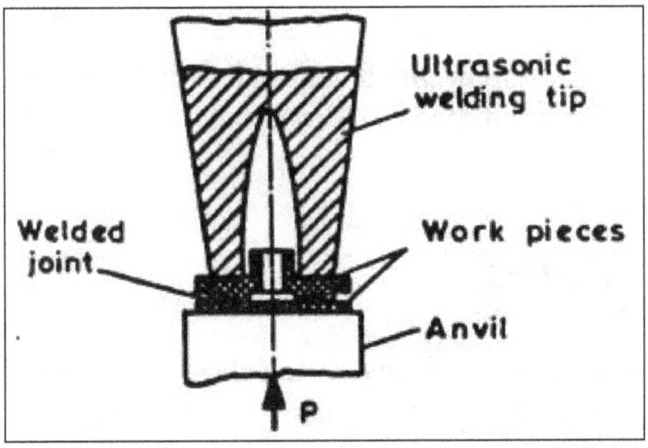

Figure: Ultrasonic welding by static-jig method.

Rigid plastics like polycarbonate with a low modulus of elasticity and a low melting point respond best to ultrasonic welding. It is also possible to weld acetal, nylon, polypropylene, high density polyethylene, acrylics, PVC, polystyrene, and synthetic textiles by ultrasonic welding. Lap and tee spot joints are made best of all. Satisfactory joints can also be made by static- jig technique, as shown in figure No edge preparation or filler metal is required for any of these joints. Dissimilar plastics can also be welded by ultrasonic welding process.

Infra-red Ray (IR) Welding

In this process welding heat supplied by an infra-red light source such as sylite glower, a chrome-steel resistance element, a quartz rod lamp, etc. To speed up the process, welding is carried out on a black backing plate from a foamed plastic, sponge rubber or thick rubberized fabric. Welding pressure is supplied by the resilience of the backup plate which is held firmly against the workpiece.

Polyethylene film can be joined satisfactorily by IR welding. Work thickness which can be welded depends upon the power of the IR source. For example, a sylite glower with a temperature of 1200 °C kept at a distance of 12 to 14 mm off the workpiece with sponge rubber backing can weld a maximum thickness of upto 2 mm. Any plastic film which can pass into viscous fluidic state and requires a low welding pressure can be welded by IR welding process. The welds produced by this process are usually free from undercuts and have high joint strength. Infra-red light can also weld sheets stacked up in a pile.

Nuclear Welding

In this process the workpieces to be welded are irradiated with a stream of neutrons.

The surfaces to be welded are given a coat of lithium or boron compound before welding. When such a coated surface is bombarded by neutrons, nuclear reaction takes place resulting in the generation of heat. The heat so produced raises the surfaces to the viscous fluidic state and therefore they can be welded. This process can be used to weld PTFE to polyethylene, polystyrene, quartz, and aluminium.

Nuclear welding has a limitation in that it cannot be applied to materials which become strongly radio-active when irradiated with neutrons.

Adhesive Bonding

Adhesive bonding is a process of joining materials in which an adhesive is placed between the faying surfaces of the components called adherends.

Adhesive bonding is similar to soldering and brazing of metals in that a metallurgical bond does not take place though the surfaces being joined may be heated but they are not melted. An adhesive may be a cement, a glue, a mucilage (sticky liquids from plants), or a paste. Though natural adhesives both of organic and inorganic origins are available, synthetic organic polymers are usually employed to adhesive bond metals.

An adhesive in the form of a liquid or a tacky solid is placed between the surfaces, to be joined, which are then mated and heat or pressure or both are applied to accomplish the joint. The advantages of adhesive bonding include bonding of dissimilar materials at low processing temperatures of 65 to 175 °C. Thin gauge materials can be bonded effectively. Adhesive joints can provide thermal and electrical insulations with smooth surface appearance resulting in uniform stress distribution.

Good vibration and sound damping can be achieved by adhesive bonding. Adhesive bonds result in significant weight saving and simplification of design.

Some adhesives can operate at temperature somewhat higher than their curing temperatures which are not possible in the case of soldered joints. However, adhesive bonds do not support high peel loads above 120 °C. Need for elaborate jigs and fixtures for assembly and curing result in high cost for equipment and tooling. Also adhesives deteriorate fast under the conditions of high humidity and temperature.

Nature of Adhesive Joints

An adhesive bond is affected by the attractive force, generally of physical nature, between an adhesive and the base material. The adhesive bond is caused either by the polar forces between the adhesive and a relatively brittle oxide film (diepole bond) or by Van der Waals forces between the adhesive and the unfilmed or clean metal.

The dipole bond is a pair of equal and opposite forces that hold two atoms together and results from a decrease in energy as two atoms are brought close together. The Van der Waal bond is defined as a secondary bond caused by the fluctuating diepole nature of an atom with all occupied electron shells filled.

When an adhesive is placed between two metal surfaces, the adhesive molecules are attracted by its neighbouring molecules as well as the metal atoms or foreign matters on the metal surfaces. If the surface energy of the adhesive is greater than that of the adherend surface, the adhesive will not wet it.

To accomplish wetting of the metal surface by the adhesive the surface energy of the metal must be greater than that of the adhesive and to achieve that it is essential to provide thorough cleanliness of the metal surfaces. Oil and grease on the surface seriously lower the surface energy of the metal surfaces and thus impair the bonding strength.

Present theory holds that adhesion is primarily due to a chemical affinity of the adhesive for the adherend and that the mechanical action, if any, is only incidental. A schematic representation of an adhesive bond is shown in figure.

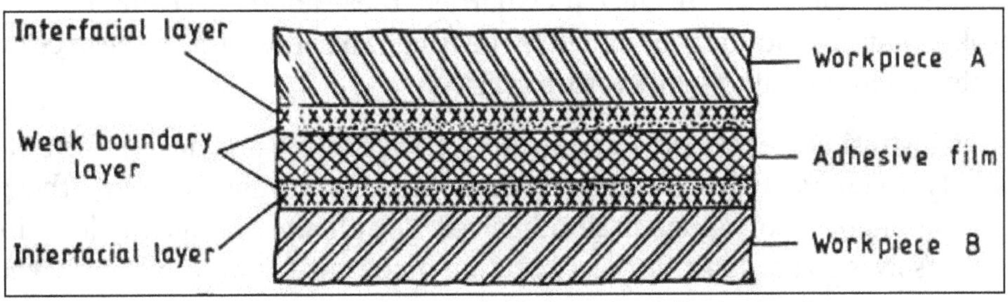

Figure: A schematic representation of an adhesive bond.

The mechanical strength of an adhesive bonded joint depends upon the joint configuration, its dimensions, the nature of adhesive and its thickness between the adhering surfaces. Generally strength of a lap joint increases with the amount of overlap (though the strength per unit area decreases), and decreases with the increasing thickness of the adhesive. The factors which may influence the strength of the joint include the contact angle between adhesive and metal, residual stress and the stress concentration in the adhesive.

Adhesives and their Classification

There are three main in gradients of most adhesives viz., a synthetic resin system, an elastomer or flexibilizer and inorganic materials.

Adhesives can be divided into two broad groups – structural adhesives and non-structural adhesives. The adhesives of the first group have high load carrying characteristics while the non-structural adhesives, also known as glues or cements, are used for low load applications, for example, waterproof latex adhesive used for tile flooring.

Structural Adhesives

Structural adhesives like plastics are classified into two groups— thermoplastic and thermosetting; the members of the former group may be re- softened repeatedly by heat though at too high a temperature, which is decided by their chemical structures, they also lose bond strength due to decomposition.

Thermoplastic Adhesives

Most commonly used thermoplastic adhesives are the polyamides, vinyls and non-vulcanizing neoprene rubber. For structural applications, the vinyls have proved very versatile, for example, polyvinyl acetate can be used to form strong bonds with metals, glass and porous materials.

Thermosetting Adhesives

Thermosetting resins are the most important materials from which metal adhesives are formed. These adhesives harden or cure by chemical reactions such as polymerisation, condensation, or vulcanization. Once they harden, these adhesives cannot be remelted and a broken joint cannot be rebounded by heating. Thermosetting adhesives are generally preferred for elevated temperature service.

Thermosetting resins are available to give strong, waterproof and heat- resistant joints. There are two general types of thermosetting structural adhesives viz., the phenolic-resin base and the epoxy-resin base adhesives. The phenol formaldehyde resins have proved themselves as amongst the best bonding materials for waterproof plywood.

Resorcinol – formaldehyde resins are similar to phenolic resins but have the advantage of getting cured at room temperature.

Epoxy resins are amongst the newest thermosetting resins and are widely acclaimed as they combine the properties of excellent action, low shrinkage, high tensile strength, toughness and chemical inertness. They can be cured at room temperature without any volatile by-products and can develop strengths between 15 to 30 MPa. Amongst the latest arrival on the scene is the 'oily metal' epoxy that bonds directly to oily metals as received with normal protective oil layer on them.

Although epoxy-based adhesives develop high shear and tensile strengths, creep and peel strengths are low. However, peel strengths of epoxy adhesives can be improved by modifying them with nylon, carboxylic functional, and nitrile copolymer rubber. Such modified epoxy adhesives can develop joint shear strength higher than 50 Mpa with high peel strength.

Other thermosetting adhesives are melamin-formaldehyde, polyurethanes, polysters, phenolic rubber, phenolic vinyl and buna, and neoprene rubbers.

Structural adhesives are also made from combinations of rubbers and synthetic resins, for example nitrile-rubber-phenolic combination can develop shear strength of 15 to 25

MPa at room temperature. These adhesives combine the strength of the phenolic resins with the flexibility and resilience of rubbers. Some of these resins can develop tensile strength of 20 to 45 MPa at room temperatures for lap joints in aluminium. Structural adhesives, developed to produce high strength are generally composed of synthetic resins or combinations of synthetic resins and elastomers. Common synthetic resins used are epoxy, urea, phenol and resorcinol.

Thermosetting adhesives are generally hard and rigid when completely cured. Elastomer resin adhesives have high strength but retain flexibility to a large extent even after curing. Flexibility of almost all adhesives can be controlled by formulation, for example, epoxy resins can be made quite flexible by modification with polysulfide rubber.

Still another class of high-temperature-resistant structural adhesives is formulated from polybenzimidazole (PBI) and polyimide (PI) which can be used successfully for the temperature range of -220 °C to 540 °C. These adhesives have been found to give excellent results for bonding aluminium, stainless steels, titanium, beryllium and reinforced plastics.

Although structural adhesives have been successfully used for bonding in aerospace applications for a number of decades, stress corrosion problems have been detected under the service conditions involving continuous or cyclic stress and a hot humid atmosphere. Adhesives cured at room temperatures degrade more rapidly in a hostile service environment than heat-cured adhesives.

Joint Design for Adhesive Bonding

The most important consideration in joint design for adhesive bonding is to know the kind of load or stress to which the part will be subjected during service. The four main types of loading encountered in such joints are shown in figure. The design must provide space enough for the adhesive to form thin bond lines in the range of 0.075 to 0.125 mm so as to achieve high bond strength.

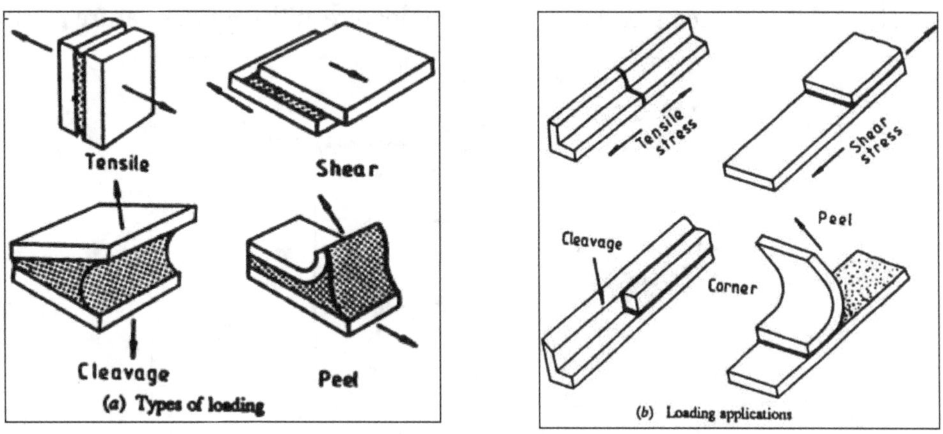

Figure: Principal types of loading encountered in adhesive bonding with actual applications.

For designing an adhesive joint three important rules are:

- The joint should preferably be subjected lo shear or tensile loading rather than cleavage or peel loading.
- The static loading of the joint should not exceed the adhesive plastic strain capacity.
- Adhesive joints subjected to low cyclic loads should be provided with sufficient overlap to minimise creep in the adhesive.

  The main types of joints employed for adhesive bonding are lap joint and the tongue and groove configuration which can be used for butt, corner or fillet joints. The mortise and tenon are used for comer joints.

Lap Joints

An adhesive joint performs best when loaded in shear as is the case in lap joints – three well-known types of which are shown in figure. In thin gauge metal bonds, joint designs can provide large bond areas; thus it is possible to produce joints that are as strong as the metal adherend.

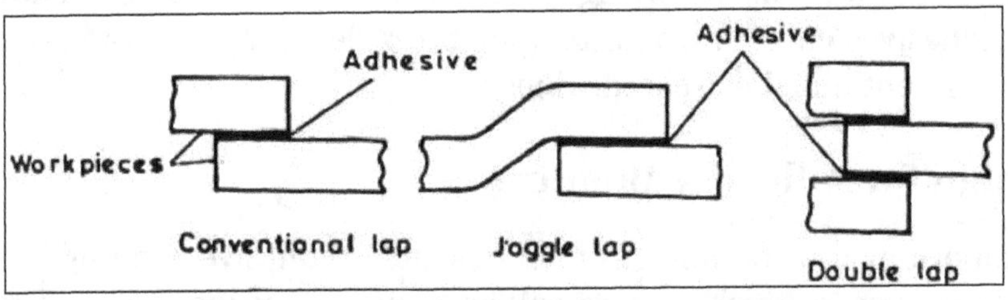

Figure: Typical lap joints for adhesive bonding.

The relationship between overlap length and joint strength, for a double shear lap joint, is shown in figure while figure shows the shear stress distribution across a lap joint caused by load P with short, medium and long overlaps. With short overlap there is uniform shear stress along the joint which can lead to creep under load resulting in premature failure.

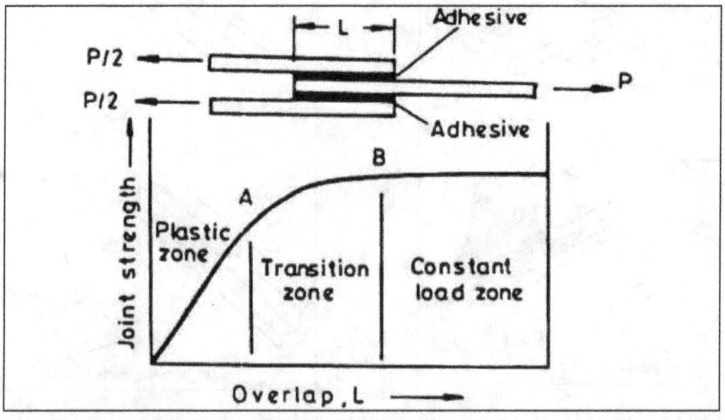

Figure: Effect of overlap length on shear strength in double-shear lap joints.

The shear stress distribution changes as the overlap length is increased so that the adhesive at the ends carries a larger portion of the load than the adhesive at the centre, thus creep potential is minimised. The joint overlap required for minimum creep depends upon the mechanical properties of the base metal, the adhesive properties and its thickness, the type of loading and the service environment.

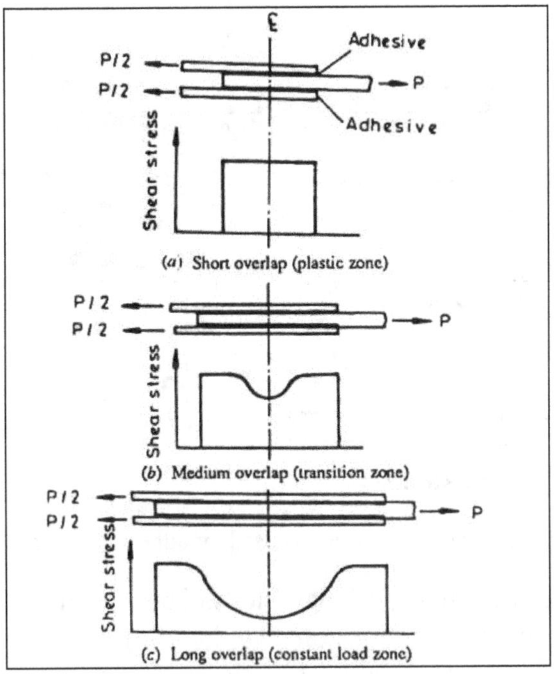

Figure: Shear stress distribution as affected by overlap for constant load.

Considerable difficulties may be encountered for designing lap joint for cleavage or peel type of loading as that causes failure to initiate at the edge of the adhesive and only a fraction of tensile load is needed to rupture bond of the same area.

The single lap joint is the most commonly used type and is adequate for many applications but the bevel lap joint provides less concentration at the edges of the bond and because thin edges of the adherends deform as the joint rotates under load that minimises the peel action.

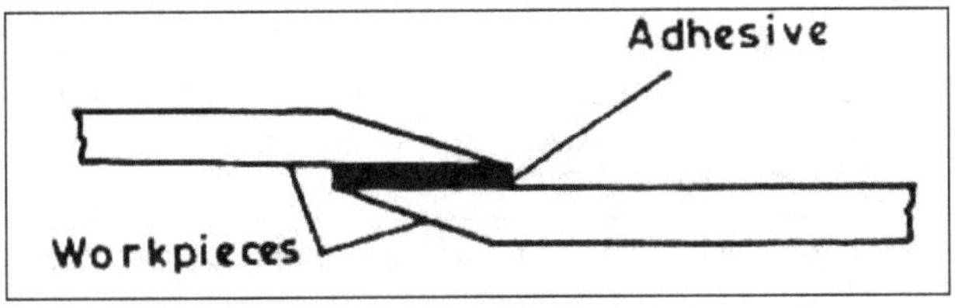

Figure: Bevelled lap joint reduces strain at edges.

When joint strength is critical and the components are thin enough to bend under load a joggled lap joint is better because the load is aligned across the joint and parallel to the band plane that minimises the possibility of cleavage loading.

Butt Joints

A square butt joint performs poorly for adhesive applications because of low effective area and high stress concentration. However, there are several ways by which the contact area between the adhesive and the adherend can be increased. These include scarf edge preparation, double butt lap, single strap, double strap, bevelled double strap and recessed double strap.

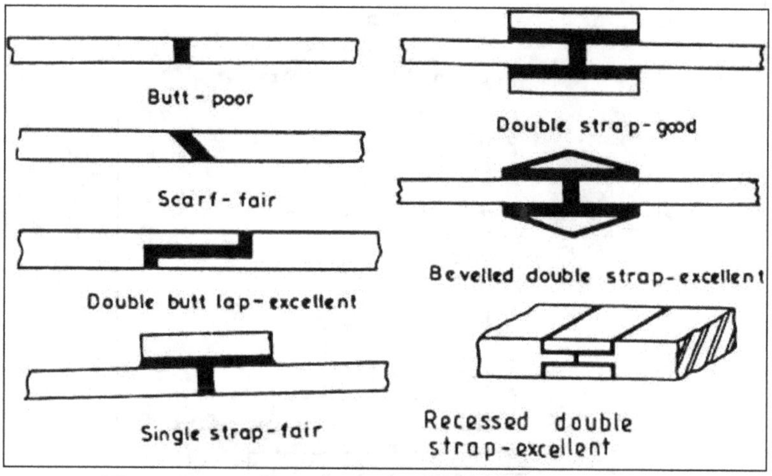

Figure: Types of butt joints used for adhesive bonding.

The tongue and groove joints not only align the load bearing interfaces with the plane of shear stress but also provide good resistance to bending. The landed-scarf tongue and groove joint is not only easy to produce but also provides configuration which aligns automatically when the parts are mated; it also controls joint length and establishes the thickness of adhesive. It is a good design in that it performs well under high compressive loads and affords a clean appearance.

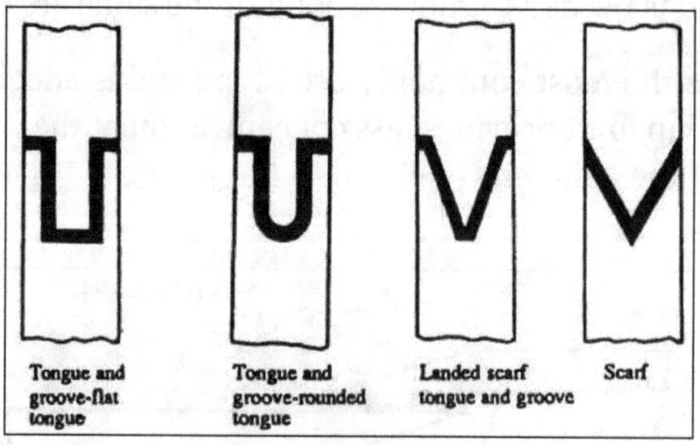

Figure: Typical tongue and groove joints used for adhesive bonding machined or extruded shapes.

Fillet or T-joint

Like the square butt joint the common T-joint may not provide adequate bending area hence various methods of improvement are adopted.

Corner Joints

The corner joints are subjected to both peel and cleavage stresses and the joint is relatively weak when the loading on a corner joint is at right angle to the adhesive. Methods of strengthening the comer joints are shown in figure.

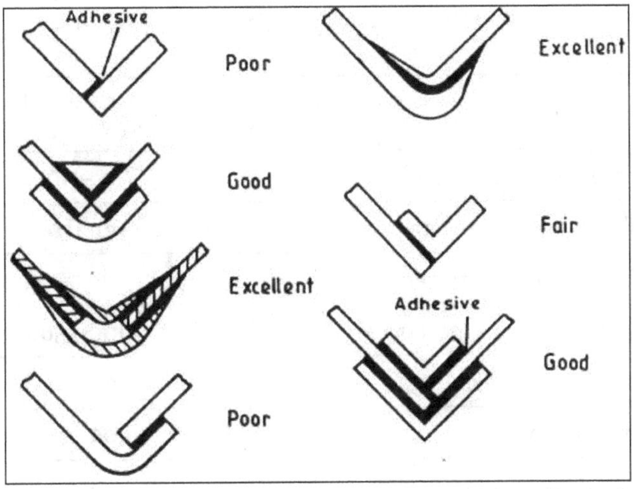

Figure: Some of the common corner joint designs used for adhesive bonding.

Tube Joints

Adhesive bonding is also used for tube joints some of which are shown in figure. Large bonded areas give strong joints with clean appearance but processing may be Complicated with some while edge preparation may be costly for some others.

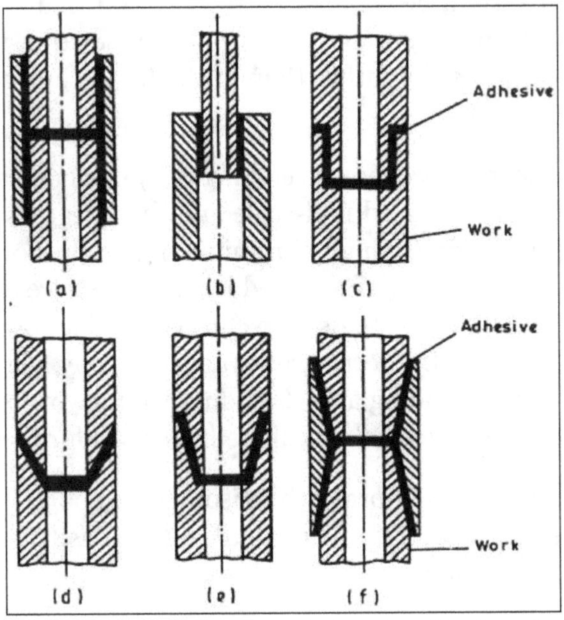

Figure: Typical tubular joints designs for adhesive joints.

Joint Strength for Adhesive Bonding

The strength developed in an adhesive joint depends upon the joint design, type of loading, service temperature, adherend material, etc. The relative shear strength for joints with different adhesives are as given in table.

Table: Relative shear strength for joints with different adhesives (after Lindbergh).

S.no.	Adhesive	Average shear strength at room temperature (N/ cm^2)
1.	Reclaimed rubber type adhesives	35-210
2.	Oil - soluble elastomers	21-140
3.	Oil - resistant elastomer	350-2100
4.	Modified epoxies (heat cured)	2100-3500
5.	Phenolic elastomer films	2100-3000
6.	Vinyl phenolic	1400-3500

Methods of Adhesive Bonding

In making adhesive joints there are essentially three steps viz., preparing the surface, applying the adhesive and curing the joint.

Preparing the Surface

Surfaces to be bonded should be cleaned by the method that ensures that the bond between the adhesive and metal surface is as strong as the adhesive itself. Failure, if it occurs, should be in the adhesive rather than at the bond- line between the adhesive and adherend.

Metal surfaces may be cleaned by chemical etching or by mechanical abrasion. Steels are first shot-blasted to remove rust and mill-scale and then degreased. Chemical etching may be necessary for preparation of high-chromium materials. For obtaining maximum strength on aluminium, surfaces are prepared by vapour degreasing and then dipped in chromic-sulphuric acid bath or anodised in chromic acid followed by careful rinse in clean water, and then air- dried. Alternatively, the metal may be roughened with abrasive to increase the effective bonding area. Grinding, filing, wire-brushing, sanding and abrasive blasting are some of the mechanical methods used for the purpose.

Certain types of plastics such as fluoro carbon isomer and polyethylene are difficult to bond and may require chemical treatment. Glass can be easily cleaned with a 30 percent hydrogen peroxide solution.

The prepared surfaces are usually tested by their affinity lo be wetted by water. It is called the water-break test. Smooth spread of water is an indication that the surface is chemically clean while the collection of droplets indicates the possibility of oil film on the surface.

To avoid the possibility of contamination of the prepared surface during storage, it is desirable to use it within a few hours. If storage is unavoidable the metal should be kept tightly wrapped or in airtight containers to minimise contamination. The etched surface must never be touched with bare hands. The operator should wear clean cotton gloves to handle the prepared surfaces as even a thumb print on an otherwise clean surface will impair adhesion.

Application of Adhesive to the Surface

Adhesives may be applied to the prepared surfaces by hand brushing, spraying, roller coating, knife-coating and dipping. They are also applied as sheet or powder, generally on a precoated surface. Sheet or tape type adhesives are gaining in popularity because there is no need for mixing and the application will be of known uniform thickness.

The thickness of the applied adhesive is referred to as 'lay-down' while the final thickness after the application of pressure and curing is called 'glue- line' thickness, for example, to achieve a glue-line thickness of 25 to 75 micron, a lay-down thickness of 0-125 to 0-375 mm of 20 percent solid wet adhesive must be applied.

The adhesive can either be applied in one thick layer on one of the parts, or in one thin layer on each of the surfaces before assembly. The latter method is preferred as it leads to a stronger bond with a longer tack life. Adhesive bonds with optimal joint strength are achieved when 0-25 to 0-75 microns of solvent-free adhesive remains after two smooth, flat, parallel surfaces are bonded together.

Lay-down thickness depends upon the porosity and smoothness of the surfaces to be bonded, the fit-up of the joint and the strength required. If the surface is porous allowance must be made in the lay-down solvent lo be absorbed by the surface, to achieve the desired glue-line thickness. Similarly, allowance must be made while coating rough surfaces so as to fill up all small depressions and attain the desired glue-line thickness; this is normally done in a single coat.

Apart from the above described general bonding procedure there are certain well established procedures for achieving optimum joint strength for specific applications. One such technique is called Redux Bonding in which the metal is first given a coat of phenol formaldehyde in a suitable solvent and then polyvinyl formaldehyde powder is scattered over the precoated surfaces before they are brought together and cured. Though polyvinyl resin is the main adhesive but precoating with phenol formaldehyde is essential to bond it to the metal. Redux Bonding is widely used, since long, for making adhesive joints for aircraft manufacture.

Assembly

Because the amount of flow for a good adhesive is very small therefore the components coated with solvent dispersed liquid adhesive should be assembled when they are tacky and wet enough to adhere to each other. The aim should be to assemble the parts when the applied adhesive is at its optimum consistency. Solvent evaporation rate may be increased by moderate heating using infra-red lamps or a hot air oven.

Provision should be made for positioning the components for mating during curing and assembly fixtures are normally used for the purpose. Care should be taken to align the parts accurately before they are mated because a strong bond is created instantly when the coated surfaces are brought together.

The assembly fixtures used for positioning should be light-weight for ease of handling. A heavy fixture is not only difficult to handle it may also act as a heat sink which can retard the heating and cooling rates during curing. The expansion rate of the fixture material should be as nearly as possible match with that of the expansion rate of assembly to minimise distortion of components and subsequent stressing of the adhesive.

Sometimes adhesive bonding is combined with resistance welding or mechanical fastening to improve the load carrying capacity of the joint. Once the parts are assembled pressure and or heat are applied to cure or set them.

Curing the Joint

With certain adhesives it is essential to apply and maintain adequate pressure during curing. The pressure should always be uniformly distributed over the entire joint. Generally, it is desirable to use as high a clamping pressure as the adherends can withstand without being crushed.

Some adhesives like epoxy can be bonded under a low pressure while some phenolic rubber adhesives require high pressures to ensure adequate flow. Normally a moderate pressure of 0-1 to 10 MPa applied in a suitable press serves the purpose well. Complex parts are placed in a plastic bag which is then evacuated allowing atmospheric pressure to apply the clamping force.

After the application of pressure, the surplus adhesive is heated through the cooling cycle preferably in an oven though electric heating pads may be listed for large components. Hydraulic platen presses are often used for applying heat and pressure to flat assemblies.

A typical curing period is 30 minutes at 145 °C though shorter times at higher temperatures may be applicable. (Heat transmitted to the adhesive being dependent upon the thermal conductivity of the adherend, the curing temperature is measured at the glueline). Curing limes may be reduced at the expense of bond strength if an accelerator is added to the adhesive.

Most of the phenolic-based structural adhesives require high curing temperatures in the range of 150 lo 205 °C for curing periods of 30 minutes to 2 hours. Some epoxies, however, can be cured at as low a temperature as 120 °C.

Extremely large components like aircraft assemblies are cured by placing them in large autoclaves. The typical operating range of such autoclaves is a pressure of upto 1-4 MPa at a maximum temperature of 175 °C. Pressure is provided by compressed air while heating is done by steam heated tubes or electrical elements.

Testing and Quality Control in Adhesive Bonding

For judging the joint quality in adhesive bonding the most commonly used destructive test is the lap shear test in which a 25 mm wide lap joint with an overlap of 12.5 mm is loaded in tension along a line parallel to the plane of the joint. Such a test is generally satisfactory for control of mixing, priming and bonding. Peel test is recommended to ascertain the adequacy of cleaning procedures; alternatively the recently developed crack extension or wedge test may be used.

The crack extension test is designed for quickly determining the durability of the adhesive joint in an environment with controlled humidity and temperature. The test specimen and the method adopted for wedging action are shown in figure. The required numbers of specimens are cut from the adhesive bonded panel.

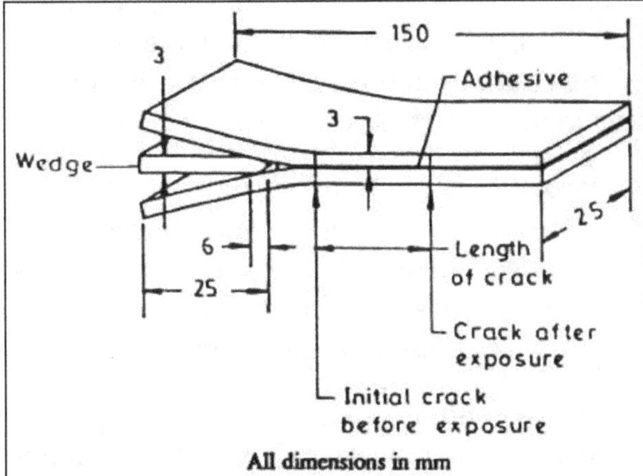

All dimensions in mm

Figure: Wedge test specimen for adhesive bonds.

The wedge is forced between the adherend at the glue- line. This separates the adhesive and produces cleavage loading at the tip opening. The location of the apex of the sheet separation is recorded. The wedged specimens are then exposed at 49 °C to an environment of 95 to 100 per cent relative humidity for 60 to 75 minutes. The distance that the apex moves during exposure is measured within two hours after exposure.

The wedge test is used for surface preparation, process control and procedures by comparing the test results with a maximum acceptable increase in adhesive crack length.

It is also used for determining the durability characteristics of the adhesive. Though the test was originally designed for adhesive bonded aluminium, it may be used for other metals with design modifications to account for differences in stiffness and yield strength.

Applications of Adhesive Bonding

Adhesive bonding of metal-to-metal accounts for less than 2% of the total metal joining applications. However, the bonding of metal to non-metals especially plastics is gaining utmost importance and is the major application of adhesive bonding.

Industries involved with aircraft and automobile construction are the major users of adhesive bonding of metals. Redux bonding was developed in early 1940's as an alternative to riveting for aircraft structures and still finds extensive use in that industry. Typical applications include fastening of stiffeners to the aircraft skin and in assembling honeycomb structures wherein honeycomb core is bonded between two sheet metal skins. Many of the joints made in the fabrication of aircraft wing and tail assemblies are by adhesive bonding; increased use is also evident in the fabrication of aircraft internal structures as well as for providing the required smooth surfaces for supersonic planes, making complex designs possible.

Adhesive bonded assemblies may comprise over 50 percent of the total area of a modern airplane. They include about 400 major assemblies including sections measuring 75 mm by 330 mm; tapered spar caps over 10 m long and panels measuring upto 1-3 m by 4-8 m. Bonded stiffeners are used on single curvature panels forming the fuselage skin. Cost of fabrication in many of these cases is reduced by 33 to 75 percent.

The main uses of adhesive bonding in automobile industry are for attaching brake lining to shoes, automatic transmission bands, and for stiffeners and fabricated box sections. Double shell panels are bonded with a high strength vinyl plastisol adhesive. Adhesive bonding reduces the number of sub-assembly details by about 50 percent, provides a smooth exterior surface, reduces noise level, and improves corrosion resistance.

Other major uses of adhesive bonding are in the fabrication of railway coaches, boats, refrigerators, storage tanks, and microwave reflectors for radar and space communications.

Safety Precautions in Adhesive Bonding

Adhesive bonding usually involves the use of corrosive materials, flammable liquids and toxic substances therefore adequate safety precautions must be observed to ensure that proper safety procedures, protective devices and protective clothing are used.

Severe skin and eye allergic reactions can result from direct contact, inhalation or ingestion of phenolics and epoxies as well as most catalysts and accelerators. It is, therefore, essential to use plastic or rubber gloves to handle potentially toxic adhesives. Eyes and face should be protected against fumes and splashing. Protective clothing should be worn at all times by those who work with the adhesives.

Adequate and effective ventilation is essential to avoid suffocation due to excessive accumulation of toxic fumes. Strict supervision is imperative to prevent inadvertent contamination of non-operating areas, for example, the contamination of door knobs, valves, handrails, etc.

References

- Metal-brazing-tips-and-advice, precious-metal-products, products-and-services: matthey.com, Retrieved 31 March, 2019
- 6-steps-to-successful-brazing, tube pipe fabrication: thefabricator.com, Retrieved 14 July, 2019
- Metal-brazing-tips-and-advice, metal-joining, precious-metal-products, products-and-services: matthey.com, Retrieved 17 May, 2019
- How-to-solder: makerspaces.com, Retrieved 19 April, 2019
- Welding-process-for-plastics-plastic-engineering, welding: yourarticlelibrary.com, Retrieved 5 February, 2019
- Adhesive-bonding-introduction-joint-design-and-methods, adhesive-bonding, welding: yourarticlelibrary.com, Retrieved 26 July, 2019

Permissions

All chapters in this book are published with permission under the Creative Commons Attribution Share Alike License or equivalent. Every chapter published in this book has been scrutinized by our experts. Their significance has been extensively debated. The topics covered herein carry significant information for a comprehensive understanding. They may even be implemented as practical applications or may be referred to as a beginning point for further studies.

We would like to thank the editorial team for lending their expertise to make the book truly unique. They have played a crucial role in the development of this book. Without their invaluable contributions this book wouldn't have been possible. They have made vital efforts to compile up to date information on the varied aspects of this subject to make this book a valuable addition to the collection of many professionals and students.

This book was conceptualized with the vision of imparting up-to-date and integrated information in this field. To ensure the same, a matchless editorial board was set up. Every individual on the board went through rigorous rounds of assessment to prove their worth. After which they invested a large part of their time researching and compiling the most relevant data for our readers.

The editorial board has been involved in producing this book since its inception. They have spent rigorous hours researching and exploring the diverse topics which have resulted in the successful publishing of this book. They have passed on their knowledge of decades through this book. To expedite this challenging task, the publisher supported the team at every step. A small team of assistant editors was also appointed to further simplify the editing procedure and attain best results for the readers.

Apart from the editorial board, the designing team has also invested a significant amount of their time in understanding the subject and creating the most relevant covers. They scrutinized every image to scout for the most suitable representation of the subject and create an appropriate cover for the book.

The publishing team has been an ardent support to the editorial, designing and production team. Their endless efforts to recruit the best for this project, has resulted in the accomplishment of this book. They are a veteran in the field of academics and their pool of knowledge is as vast as their experience in printing. Their expertise and guidance has proved useful at every step. Their uncompromising quality standards have made this book an exceptional effort. Their encouragement from time to time has been an inspiration for everyone.

The publisher and the editorial board hope that this book will prove to be a valuable piece of knowledge for students, practitioners and scholars across the globe.

Index

www.ingramcontent.com/pod-product-compliance
Lightning Source LLC
LaVergne TN
LVHW080607210926
848485LV00003B/118